Lecture Notes in Control and Information Sciences

Edited by A. V. Balakrishnan and M. Thoma

Vol. 1: Distributed Parameter Systems: Modelling and Identification
Proceedings of the IFIP Working Conference, Rome, Italy, June 21–26, 1976
Edited by A. Ruberti
V, 458 pages. 1978

Vol. 2: New Trends in Systems Analysis
International Symposium, Versailles, December 13–17, 1976
Edited by A. Bensoussan and J. L. Lions
VII, 759 pages. 1977

Vol. 3: Differential Games and Applications
Proceedings of a Workshop, Enschede, Netherlands, March 16–25, 1977
Edited by P. Hagedorn, H. W. Knobloch, and G. J. Olsder
XII, 236 pages. 1977

Vol. 4: M. A. Crane, A. J. Lemoine
An Introduction to the Regenerative Method for Simulation Analysis
VII, 111 pages. 1977

Vol. 5: David J. Clements, Brian D. O. Anderson
Singular Optimal Control: The Linear Quadratic Problem
V, 93 pages. 1978

Vol. 6: Optimization Techniques
Proceedings of the 8th IFIP Conference on Optimization Techniques, Würzburg, September 5–9, 1977
Part 1
Edited by J. Stoer
XIII, 528 pages. 1978

Vol. 7: Optimization Techniques
Proceedings of the 8th IFIP Conference on Optimization Techniques, Würzburg, September 5–9, 1977
Part 2
Edited by J. Stoer
XIII, 512 pages. 1978

Vol. 8: R. F. Curtain, A. J. Pritchard
Infinite Dimensional Linear Systems Theory
VII, 298 pages. 1978

Vol. 9: Y. M. El-Fattah, C. Foulard
Learning Systems:
Decision, Simulation, and Control
VII, 119 pages. 1978

Vol. 10: J. M. Maciejowski
The Modelling of Systems with Small Observation Sets
VII, 241 pages. 1978

Vol. 11: Y. Sewaragi, T. Soeda, S. Omatu
Modelling, Estimation, and Their Applications for Distributed Parameter Systems
VI, 269 pages. 1978

Vol. 12: I. Postlethwaite, A. G. J. McFarlane
A Complex Variable Approach to the Analysis of Linear Multivariable Feedback Systems
IV, 177 pages. 1979

Vol. 13: E. D. Sontag
Polynomial Response Maps
VIII, 168 pages. 1979

Vol. 14: International Symposium on Systems Optimization and Analysis
Rocquentcourt, December 11–13, 1978;
IRIA LABORIA
Edited by A. Bensoussan and J. Lions
VIII, 332 pages. 1979

Vol. 15: Semi-Infinite Programming
Proceedings of a Workshop, Bad Honnef, August 30 – September 1, 1978
V, 180 pages. 1979

Vol. 16: Stochastic Control Theory and Stochastic Differential Systems
Proceedings of a Workshop of the „Sonderforschungsbereich 72 der Deutschen Forschungsgemeinschaft an der Universität Bonn"
which took place in January 1979 at Bad Honnef
VIII, 615 pages. 1979

Vol. 17: O. I. Franksen, P. Falster, F. J. Evans
Qualitative Aspects of Large Scale Systems
Developing Design Rules Using APL
XII, 119 pages. 1979

Vol. 18: Modelling and Optimization of Complex Systems
Proceedings of the IFIP-TC 7 Working Conference
Novosibirsk, USSR, 3–9 July, 1978
Edited by G. I. Marchuk
VI, 293 pages. 1979

Vol. 19: Global and Large Scale System Models
Proceedings of the Center for Advanced Studies (CAS)
International Summer Seminar
Dubrovnik, Yugoslavia, August 21–26, 1978
Edited by B. Lazarević
VIII, 232 pages. 1979

Vol. 20: B. Egardt
Stability of Adaptive Controllers
V, 158 pages. 1979

Vol. 21: Martin B. Zarrop
Optimal Experiment Design for Dynamic System Identification
X, 197 pages. 1979

For further listing of published volumes please turn over to inside of back cover.

Lecture Notes in Control and Information Sciences

Edited by A.V. Balakrishnan and M. Thoma

31

Berc Rustem

Projection Methods in Constrained Optimisation and Applications to Optimal Policy Decisions

Springer-Verlag Berlin Heidelberg GmbH 1981

Author
Dr. Berc Rustem
Control Section
Dept. of Electrical Engineering
Imperial College of Science and Technology
Exhibition Road
London. SW7 2BT – England

ISBN 978-3-540-10646-3 ISBN 978-3-540-38603-2 (eBook)
DOI 10.1007/978-3-540-38603-2

Originally published by Springer-Verlag Berlin Heidelberg New York in 1981

ABSTRACT

This work is concerned with projection methods in constrained optimisation and the application of projection techniques to policy optimisation problems. The constrained optimisation problem of minimising a nonlinear function of n variables subject to equality and inequality constraints is also known as the nonlinear programming problem. Projection methods for constrained minimisation involve projections of descent directions. The basic idea underlying these methods is the principle of projections which may be considered to be the generalisation of the fact that in n-dimensional Euclidean space the shortest vector from a point to a subspace is orthogonal to the subspace.

In effect, this work consists of two parts. The first part, Chapters 1 - 4, is concerned with the development of projection techniques for different aspects of constrained optimisation. Chapter 1 provides a unified approach to the derivation of projection techniques and reviews existing methods. The application of projection techniques to the computation of a feasible point of a linearly constrained region is discussed in Chapter 2. In Chapter 3 a projection method is discussed for the quadratic programming problem of minimising a quadratic objective function subject to inequality constraints. This method requires an initial feasible point. In Chapter 4 a method is discussed for solving the nonlinear programming problem. This method requires solutions to quadratic minimisation subproblems.

The second part of the work is concerned with the application of projection techniques to computational problems in policy optimisation. A fundamental problem in the optimisation of policy decisions is the specification of a suitable objective function. In Chapter 5 an iterative method is given for specifying objective functions. In Chapter 6 policy optimisation algorithms, based on extensions of projection methods in Chapter 1 , are discussed. These algorithms minimise quadratic objective functions subject to large nonlinear models treated as equality constraints.

ACKNOWLEDGEMENTS

I would like to thank Dr. Heather Liddell for her valuable guidance and great patience.

I am also grateful to Dr. Alan Frieze for helpful discussions and valuable comments and to Drs. Kumaraswamy Velupillai, Sean Holly for providing the economics background of Chapter 5.

Finally, I wish to thank Miss Claudia Schamberger for patiently typing the manuscript.

CONTENTS

CONVENTIONS AND SYMBOLS

Assumptions, Definitions, Lemmas, Theorems etc. are given a number preceded by the chapter number. Equations and algorithms are given a number, preceded by the section and chapter numbers in which they occur.

The end of a proof or a particular train of thought is denoted by □ . The symbol $\triangleq$ denotes 'defined equal to'.

Since each chapter deals with a different aspect of constrained optimisation (e.g. feasible point algorithms, quadratic programming, nonlinear programming, etc.) the symbols used in each chapter have been slightly adapted to the needs of that chapter. The usage of commonly used symbols is given below.

Symbol

$\underline{a}$	coefficient vector of the linear term of the quadratic function
$\underline{b}$	right hand side vector of a system of linear equalities and/or inequalities
$\underline{d}$	descent direction
E^n	Euclidean n-space
F	policy maker's admissible region for $\underline{U}$ and $\underline{Y}$
$\underline{F}(\underline{Y},\underline{U}) = \underline{0}$	Equations of the econometric model
$f(\underline{x})$	nonlinear objective function, $\underline{x} \in E^n$

G	Hessian matrix
$G(\underline{U})$	reduced form of $J(\underline{Y},\underline{U})$ with $\underline{Y}$ eliminated
$\underline{g}(\underline{U})$	reduced form equations of the econometric model
$\underline{g}(\underline{x})$	vector of inequality and/or equality constraints
H	inverse Hessian matrix
$\underline{h}(\underline{x})$	vector of equality constraints
$I(\underline{x})$	index set of active constraints at $\underline{x}$
$J(\underline{Y},\underline{U})$	quadratic performance index for optimal control
$L(\underline{x},\underline{\lambda})$	Lagrangian function
N	matrix of constraint normals
$P, P[H], \bar{P}, \hat{P}$	projection operators
$q, \bar{q}$	quadratic functions approximating $f(\underline{x})$
$\hat{q}$	$\bar{q}$ with approximate Hessian
R	the feasible region
$S(\underline{x})$	index set of the satisfied constraints at $\underline{x}$
$\underline{U}$	vector of controls (policy instruments)
$V(\underline{x})$	index set of the violated constraints at $\underline{x}$
$\underline{x}_u$	the point reached along an unconstrained direction
$\underline{x}_p$	the projection of $\underline{x}_u$
$\underline{Y}$	vector of outputs (endogenous variables)

α	steplength factor along an unconstrained direction
$\underline{\gamma}$	change in the gradient vector of the objective function
$\underline{\theta}(\underline{x})$	residual vector of a system of linear equalities and/or inequalities evaluated at $\underline{x}$
$\underline{\lambda}$	the multiplier vector
τ	steplength along a projected direction
$\nabla_z(\cdot)$	gradient vector of (.) with respect to z (z is omitted wherever the argument is obvious)

CHAPTER 1

NONLINEARLY CONSTRAINED OPTIMISATION TECHNIQUES BASED ON PROJECTIONS

1.1 INTRODUCTION

This work is concerned with projection methods for constrained optimisation and the application of projection techniques to policy optimisation problems.

The principle of projections may be considered to be the generalisation of the fact that in n-dimensional Euclidean space the shortest vector from a point to a subspace is orthogonal to the subspace. Projections have a wide application in mathematics and especially in systems theory. For example, in the analysis of distributed systems, the method of orthogonal projections is applied by Mikhlin (1964) to solve the Dirichlet problem. Another important application arises in the derivation of the least-squares, minimum variance and recursive estimation (Kalman filtering) techniques in statistical estimation (see, e.g. Davis (1977), Luenberger (1969, Chapter 4)). Such techniques have found wide application in engineering and econometrics (see, e.g. Rustem and Velupillai (1978, 1979)). In all these applications a suitable normed linear vector space such as a Hilbert or Euclidean space is chosen and the projection of vectors in this space onto a a subspace is considered.

The projection problem involves the determination of the shortest vector to a subspace and is therefore inherently an optimisation problem. With the advent of nonlinear constrained optimisation techniques the idea of formulating optimisation problems as projection problems was introduced by Rosen (1960). In subsequent sections different implementations of this idea are discussed. In section (1.1.1) fundamental concepts and results in constrained optimisation are given. In Section (1.1.2) a unified approach to projection operators is discussed. Sections (1.1.3) and (1.1.4) consider methods using projection operators based on complementary subspaces. In Section (1.2.1) extensions for solving nonlinearly constrained optimisation problems are discussed. In Section (1.2.2) active set strategies are reviewed. In Section (1.3) the methods discussed in this work are reviewed.

1.1.1 Fundamental Concepts and Results

The nonlinear constrained optimisation problem

$$\min \{f(\underline{x}) \mid \underline{g}(\underline{x}) \geq \underline{o} \ , \ \underline{h}(\underline{x}) = \underline{o} \} \tag{1.1.1}$$

is also known as the nonlinear programming problem. Let E^n denote the n-dimensional Euclidian space, $\underline{x} \in E^n$ and let the mappings

$$f : E^n \to E^1 \tag{1.1.2,a}$$

$$\underline{g} : E^n \to E^{m_o} \tag{1.1.2,b}$$

$$\underline{h} : E^n \to E^r \tag{1.1.2,c}$$

be continuously differentiable functions. Furthermore, $f(\underline{x})$ is called the objective function,

$$\underline{g}(\underline{x}) \triangleq (g_1(\underline{x}), \ldots, g_{m_o}(\underline{x}))^T \tag{1.1.3}$$

is the m_o-dimensional function corresponding to the inequality constraints, and

$$\underline{h}(\underline{x}) \triangleq (h_1(\underline{x}), \ldots, h_r(\underline{x}))^T \tag{1.1.4}$$

is the r-dimensional function corresponding to the equality constraints in (1.1.1).

A feasible point of problem (1.1.1) is a point $\underline{x}$ that belongs to the feasible region R defined as

$$R \triangleq \{\underline{x} \in E^n \mid \underline{g}(\underline{x}) \geq \underline{o} \ , \ \underline{h}(\underline{x}) = \underline{o}\} \ . \tag{1.1.5}$$

An element, $g_i(\underline{x})$, of $\underline{g}(\underline{x})$ is said to be active when strict equality, $g_i(\underline{x}) = o$, holds for that element.

A global minimum of problem (1.1.1) is defined as a point $\underline{x}^*$ for which

$$f(\underline{x}^*) < f(\underline{x}) \ , \ \forall \ \underline{x} \in R \qquad (1.1.6)$$

holds. A strong local minimum however, is the point $\underline{x}_c$ for which $f(\underline{x}_c)$ is strictly lower over R in a specified neighbourhood, i.e.

$$f(\underline{x}_c) < f(\underline{x}) \qquad \forall \underline{x} \in \{\underline{x} \in R \,|\, \|\underline{x} - \underline{x}_c\| < \delta, \ \underline{x} \neq \underline{x}_c\} \qquad (1.1.7)$$

for some scalar $\delta > o$ determining the neighbourhood. In this chapter the determination of such local optima for (1.1.1) is discussed.

The conditions for $\underline{x}_c$ to be a strong local minimum of (1.1.1) may be given by first introducing the Lagrangian function for (1.1.1)

$$L(\underline{x},\underline{\lambda},\underline{\mu}) \triangleq f(\underline{x}) + \langle \underline{h}(\underline{x}),\underline{\mu} \rangle - \langle \underline{g}(\underline{x}),\underline{\lambda} \rangle \qquad (1.1.8)$$

where

$$\underline{\lambda} \triangleq (\lambda_1, \ldots, \lambda_{m_o})^T \in E_+^{m_o} \triangleq \{\underline{\lambda} \in E^{m_o} | \underline{\lambda} \geq \underline{o}\} \qquad (1.1.9)$$

and

$$\underline{\mu} \triangleq (\mu_1, \ldots, \mu_r)^T \in E^r \qquad (1.1.10)$$

are known as the Kuhn-Tucker or Generalized Lagrange Multipliers. The first order necessary conditions for a strong local minimum are that there exist vectors $\underline{x}_c, \underline{\lambda}_c, \underline{\mu}_c$ which satisfy

$$\nabla L(\underline{x}_c, \underline{\lambda}_c, \underline{\mu}_c) = \nabla f(\underline{x}_c) + \nabla \underline{h}(\underline{x}_c)\underline{\mu}_c - N_{m_o}(\underline{x}_c)\underline{\lambda}_c = \underline{o} \qquad (1.1.11)$$

$$\underline{h}(\underline{x}) = \underline{o}, \qquad (1.1.12)$$

$$\lambda_c^i \, g_i \, (\underline{x}_c) = o \, , \qquad i = 1, \ldots, m_o \qquad (1.1.13)$$

$$\underline{g} \, (\underline{x}_c) \geq \underline{o} \qquad (1.1.14)$$

$$\underline{\lambda}_c \geq \underline{o} \qquad (1.1.15)$$

provided f, $\underline{g}$ and $\underline{h}$ are at least once continuously differentiable at $\underline{x}_c$ and that those columns of the $(n \times m_o)$ matrix (N_{m_o}),

$$N_{m_o}(\underline{x}) \triangleq [\nabla g_1(\underline{x}), \ldots, \nabla g_{m_o}(\underline{x})], \qquad (1.1.16)$$

corresponding to the active elements of $\underline{g}$, at $\underline{x}_c$, and the columns of the $(n \times r)$ matrix $\nabla h(\underline{x})$ defined as

$$\nabla \underline{h}(\underline{x}) \triangleq [\nabla h_1(\underline{x}), \ldots, \nabla h_r(\underline{x})] \qquad (1.1.17)$$

are linearly independent at $\underline{x}_c$ (see, Luenberger (1973, pp. 232-233)).

For twice continuously differentiable f, $\underline{g}$, and $\underline{h}$ at x_c, the second order sufficiency conditions for a strong local minimum of (1.1.1) are given by (1.1.11) - (1.1.15) and

$$< \underline{v}, \nabla^2 L(\underline{x}_c, \underline{\lambda}_c, \underline{\mu}_c) \, \underline{v} > \; > o \qquad (1.1.18)$$

for all nonzero n-vectors $\underline{v}$ on the subspace

$$R' \triangleq \{\underline{v} | \nabla \underline{h}(\underline{x}_c)^T \, \underline{v} = \underline{o} \, , < \nabla g_j(\underline{x}_c), \underline{v} > = o \quad \forall_j \in J\} \qquad (1.1.19)$$

where $\nabla^2 L(\underline{x}, \underline{\lambda}, \underline{\mu})$ in (1.1.18) is the Hessian matrix of (1.1.8) with respect to $\underline{x}$ and

$$J \triangleq \{j | \; j \in I \quad , \lambda_c^j > o\} \qquad (1.1.20)$$

$$I \triangleq \{i | \; g_i(\underline{x}_c) = o\} \qquad (1.1.21)$$

(see Luenberger (1973, pp. 234 - 235)).

The conditions (1.1.11) to (1.1.15) together with the condition (1.1.18) for the Hessian matrix of the Lagrangian (1.1.8) to be positive definite on R′ are the necessary and sufficient conditions for a strong local minimum.

The equality constraints $\underline{h}(\underline{x}) = \underline{o}$ will, in general, be treated as a special case of the inequality constraints. Accordingly, the feasible region R may be redefined as

$$R \triangleq \{\underline{x} \in E^n \mid \underline{g}(\underline{x}) \geq \underline{o}\}. \qquad (1.1.22)$$

If the elements $g_i(\underline{x})$ of the vector $\underline{g}(\underline{x})$ are concave functions, then R is a convex set. All equality constraints represented in $\underline{g}(\underline{x})$ must be linear for R to be a convex set since by definition for $g_i(\underline{x})$ to be concave $-g_i(\underline{x})$ must be convex. With R convex if the objective function is also assumed to be convex then every local minimum x_c to the problem

$$\min \{f(\underline{x}) \mid \underline{x} \in R\} \qquad (1.1.23)$$

is also a global minimum (Fiacco & McCormick (1968, Theorem 18)). Under these assumptions of convexity, the Kuhn-Tucker necessary conditions for (1.1.23)

$$\nabla f(\underline{x}_c) - N_{m_o}(\underline{x}_c)\ \underline{\lambda}_c = 0 \qquad (1.1.24)$$

$$\underline{g}(\underline{x}_c) \geq \underline{o} \qquad (1.1.25)$$

$$\langle \underline{\lambda}_c, \underline{g}(\underline{x}_c) \rangle = 0 \qquad (1.1.26)$$

$$\underline{\lambda}_c \geq \underline{o} \qquad (1.1.27)$$

are also the sufficient conditions for $\underline{x}_c$ to be a constrained minimum of (1.1.23) (Fiacco & McCormick (1968, Theorem 19)).

To locate a strong local constrained minimum of the constrained optimisation problem (1.1.1) a sequence of subproblems are devised to generate a sequence of points converging to the required minimum. Projection methods are usually based on generating a sequence of quadratic programming problems (see Chapter 3) making local quadratic approximations to the objective function and linear approximations to the constraints. In subsequent sections different implementations of this approach will be discussed.

1.1.2 Projection Methods for Constrained Minimasation

Most projection methods reduce the inequality constrained problem (1.1.23) to a sequence of equality constrained subproblems assuming that some of the inequality constraints (1.1.22) are satisfied as equalities for that subproblem. The process of selecting the constraints which are satisfied as equalities in these subproblems is discussed in Section (1.2.2). The following assumptions are necessary for the derivation of the basic relationships utilised in projection algorithms.

Assumption (1.1)

At $\underline{x}_d$, the objective function, $f(\underline{x}_d)$, may be approximated by a quadratic function such that

$$f(\underline{x}_d) = q(\underline{d}) + 0\ (\|\underline{d}\|^3) \qquad (1.1.28)$$

where

$$\underline{x}_d = \underline{x}_k + \underline{d} \qquad (1.1.29)$$

and

$$q(\underline{d}) = f(\underline{x}_k) + < \nabla f(\underline{x}_k), \underline{d} > + \tfrac{1}{2} < \underline{d}, G_k \underline{d} > . \qquad (1.1.30)$$

The quadratic $q(\underline{d})$ is a second order Taylor expansion of $f(\underline{x})$ about $\underline{x}_k$ and G_k is the Hessian matrix of $f(\underline{x})$ evaluated at $\underline{x}_k$. It is further assumed that G_k is nonsingular unless otherwise stated.

Assumption (1.2)

The constraints $\underline{g}(\underline{x}) \geq \underline{o}$ are assumed to be linear and of the form

$$< \underline{n}_i , \underline{x} > - b_i \geq o \quad , \quad i = 1, \ldots, m_o \qquad (1.1.31)$$

or written in matrix form

$$N_{m_o}^T \underline{x} - \underline{b}_{m_o} \geq \underline{o} \qquad (1.1.32)$$

where the vectors $\underline{n}_i$ are the columns of the matrix N_{m_o} and $\underline{b}_{m_o}$ is an m_o-dimensional vector. A number, m, of these constraints may be satisfied as equalities, at $\underline{x}$, i.e.

$$N_m^T \underline{x} - \underline{b}_m = \underline{o} \; , \qquad (1.1.33)$$

where the columns of N_m are assumed to be linearly independent. Clearly, when $\underline{x}$ satisfies all the constraints (1.1.31) as strict inequalities, the integer $m=o$ and $\underline{x}$ lies inside the feasible region R, (1.1.22).

Definition (1.1)

The n-m dimensional subspace $\Omega_o \subset E^n$ is defined as

$$\Omega_o \triangleq \{ \underline{d}_1 \in E^n \mid N_m^T \underline{d}_1 = \underline{o} \} . \qquad (1.1.34)$$

The linear manifold associated with Ω_o is denoted as Ω ,

$$\Omega \triangleq \{ \underline{x} \in E^n \mid N_m^T \underline{x} = \underline{b}_m \} . \qquad (1.1.35)$$

Definition (1.2)

The m-dimensional subspace $\bar{\Omega} \subset E^n$ is defined such that

$$\forall \underline{d}_2 \in \bar{\Omega} \quad , \quad \underline{d}_2 = H_k N_m \underline{\lambda} \,, \qquad (1.1.36)$$

where $\underline{\lambda}$ is an m-dimensional vector and $H_k = G_k^{-1}$.

Thus $\forall \underline{d}_1 \in \Omega_o$ and $\forall \underline{d}_2 \in \bar{\Omega}$ it is verified

that $< \underline{d}_1, G_k \underline{d}_2 > = o$.

And for $\underline{d} \in E^n$ there exists a representation

$$\underline{d} = \underline{d}_1 + \underline{d}_2 \qquad (1.1.37)$$

since $E^n = \Omega_o \oplus \bar{\Omega}$.

Newton's method for the unconstrained optimisation of $f(\underline{x})$ is based on the generation of the descent direction which finds the unconstrained optimum of the quadratic approximation (1.1.30) to the objective function (1.1.28). For the linearly constrained case the descent direction thus generated will also be required to satisfy $N_m^T \underline{d}_1 = \underline{o}$ where the columns of N_m correspond to the constraints satisfied as equalities at $\underline{x}_k$. Thus $\underline{d}_1$ is generated by the solution of

$$\min \; \{ q(\underline{d}_1) \mid \underline{d}_1 \in \Omega_o \}. \qquad (1.1.38).$$

The successive solutions of (1.1.38) generate a sequence of directions that lead to a solution, $\underline{x}_c$, of the linear equality constrained problem

$$\min \quad \{f(\underline{x}) \mid \underline{x} \in \Omega\} \cdot \qquad (1.1.39)$$

Associated with the point $\underline{x}_c$ are the corresponding Lagrange multipliers which satisfy

$$N_m \underline{\lambda}_c = \nabla f(\underline{x}_c). \qquad (1.1.40)$$

This system of n equations in m unknowns has the unique solution

$$\underline{\lambda}_c = (N_m^T N_m)^{-1} N_m^T \nabla f(\underline{x}_c) \qquad (1.1.41)$$

since $\nabla f(\underline{x}_c)$ lies in the range of N_m (Powell (1974,b, Theorem 1.5)). At points other than $\underline{x}_c$ there is no vector $\underline{\lambda}_c$ which satisfies (1.1.40). Only estimates of $\underline{\lambda}_c$ may be obtained and the solution of (1.1.38) yields such estimates.

There are two alternative ways of solving (1.1.38). To explain these clearly it is necessary to note that the solution of (1.1.38) is the projection of the unconstrained minimum of (1.1.30) onto Ω_o. This will be proved in Theorem (3.1).

A useful definition of the projection $\underline{d}_1^*$ of the vector $\underline{d} \in E^n$ onto the subspace Ω_o is given by

$$\underline{d}_1^* = \arg\min\{\|\underline{d} - \underline{d}_1\|^2_{G_k} \mid \underline{d}_1 \in \Omega_o\} \qquad (1.1.42)$$

where $\|\underline{d} - \underline{d}_1\|^2_{G_k} = \langle \underline{d} - \underline{d}_1 , G_k(\underline{d} - \underline{d}_1) \rangle$.

The unconstrained minimum of (1.1.30) is, with $H_k = G_k^{-1}$,

$$\underline{d} = -H_k \nabla f(\underline{x}_k) \qquad (1.1.43)$$

which is also Newton's direction for the unconstrained minimisation of $f(\underline{x})$. The projection of $\underline{d}$ onto Ω_o is therefore given by

$$\min \{\| -H_k \nabla f(\underline{x}_k) - \underline{d}_1\|^2_{G_k} \mid \underline{d}_1 \in \Omega_o\}. \qquad (1.1.44)$$

The first method for computing $\underline{d}_1^*$ establishes the n-m basis vectors $\underline{Z}_j$, j=1, ..., n-m, of subspace Ω_0 such that any vector in Ω_0 may be expressed as a linear combination of these basis vectors. Thus, defining the nx(n-m) matrix

$$Z \triangleq (\underline{Z}_1, \ldots, \underline{Z}_{n-m}) \tag{1.1.45}$$

all vectors $\underline{d}_1 \in \Omega_0$ may be expressed as

$$\underline{d}_1 = Z\underline{v} \tag{1.1.46}$$

where $\underline{v}$ is an n-m vector. The problem (1.1.44) may be expressed as an unconstrained minimisation problem in n-m dimensions,

$$\min_{\underline{v}} \{\| -H_k \nabla f(\underline{x}_k) - Z\underline{v} \|^2_{G_k}\} \tag{1.1.47}$$

the solution of which may be written as

$$\nabla_{\underline{v}} \| -H_k \nabla f(\underline{x}_k) - Z\underline{v} \|^2_{G_k} = 2\, Z^T(-\nabla f(\underline{x}_k) - G_k Z\underline{v})$$

$$= \underline{0}$$

thus,

$$(Z^T G_k Z)\, \underline{v} = -Z^T \nabla f(\underline{x}_k) \tag{1.1.48}$$

and using (1.1.46)

$$\underline{d}_1^* = -Z(Z^T G_k Z)^{-1}\, Z^T \nabla f(\underline{x}_k). \tag{1.1.49}$$

This method requires only $(Z^T G_k Z)$ to be positive definite and not G_k itself. Clearly, this could be an advantage when the original matrix G_k is not positive definite and a positive definite approximation to it is being used. This approximation to G_k is in E^n , whereas a positive definite approximation to a non positive definite $(Z^T G_k Z)$ is made in the n-m dimensional subspace $\Omega_0 \subset E^n$. Gill and Murray (1974,a) give an example to illustrate that such an approximation which is undesirable in E^n becomes acceptable in Ω_o.

The second method for solving (1.1.38) is based on using (1.1.37) such that

$$\underline{d}_1^* = \underline{d} - \underline{d}_2^* \qquad (1.1.50)$$

where $\underline{d}_2^*$ is the projection of $\underline{d}$ onto the subspace $\bar{\Omega}$. Since the vectors in this subspace may be expressed as $H_k N_m \underline{\lambda}$ for some m-vector $\underline{\lambda}$, the projection of $\underline{d}$ onto $\bar{\Omega}$ is that vector $\underline{d}_2 \in \bar{\Omega}$ given by

$$\underline{d}_2^* = \arg\min \{ \| \underline{d} - \underline{d}_2 \|^2_{G_k} \mid \underline{d}_2 \in \bar{\Omega} \} .$$

In turn, the above minimisation is solved by

$$\underline{\lambda}^* = \arg \min_{\underline{\lambda}} \{ \| \underline{d} - H_k N_m \underline{\lambda} \|^2_{G_k} \} \qquad (1.1.51)$$

and thus $\underline{d}_2^* = HN\underline{\lambda}^*$.

If $\underline{d}$ is given by (1.1.43) then the unconstrained minimum of (1.1.51) with respect to $\underline{\lambda}$ is given by the condition

$$\nabla_{\underline{\lambda}} \| -H_k \nabla f(\underline{x}_k) - H_k N_m \underline{\lambda} \|^2_{G_k} = 2N_m^T(H_k \nabla f(\underline{x}_k) + H_k N_m \underline{\lambda})$$

$$= \underline{0}.$$

This yields

$$\underline{\lambda}^* = -(N_m^T H_k N_m)^{-1} N_m^T H_k \nabla f(\underline{x}_k). \qquad (1.1.52)$$

Using (1.1.52) with (1.1.50), the expression for $\underline{d}_1^*$ can be written as

$$\underline{d}_1^* = \underline{d} - H_k N_m \underline{\lambda}^* \qquad (1.1.53)$$

$$= -H_k \nabla f(\underline{x}_k) + H_k N_m (N_m^T H_k N_m)^{-1} N_m^T H_k \nabla f(\underline{x}_k)$$

$$= -(I - H_k N_m (N_m^T H_k N_m)^{-1} N_m^T) H_k \nabla f(\underline{x}_k).$$

(1.1.54)

To show that $\underline{d}_1^*$ computed by (1.1.49) and (1.1.54) are identical if $N_m^T H_k N_m$ and $Z^T G_k Z$ are positive definite, consider the relationship

$$N_m^T \underline{d}_1^* = \underline{o} \tag{1.1.55}$$

which follows from Definition (1.1). Thus, (1.1.54) may be written in the form (1.1.46),

$$(I - H_k N_m (N_m^T H_k N_m)^{-1} N_m^T) H_k \nabla f(\underline{x}_k) = Z\underline{v}, \tag{1.1.56}$$

for some n-m vector $\underline{v}$. Premultiplying (1.1.56) by $Z^T G_k$ the expression

$$Z^T \nabla f(\underline{x}_k) = Z^T G_k Z \underline{v}$$

is obtained since it follows from (1.1.34) and (1.1.36) in Definitions (1.1) - (1.2) that $Z^T N_m = o$. Thus the expressions (1.1.49) and (1.1.54) are identical.

A condition for $\underline{x}_c$ to be a solution of the minimisation problem with linear equality constraints is that $\underline{d}_1^*$ given by (1.1.54) is zero at $\underline{x}_c$. Alternatively, since $\nabla f(\underline{x}_c) = N_m \underline{\lambda}_c$ and $Z^T N_m = o$, this condition can be written as $Z^T \nabla f(\underline{x}_c) = \underline{o}$.

Since for $\underline{x}_k \neq \underline{x}_c$ there is no vector $\underline{\lambda}_k$ for which

$$N_m \underline{\lambda}_k = \nabla f(\underline{x}_k) \tag{1.1.57}$$

holds, a simple estimate of the Lagrange multipliers at $\underline{x}_k$ is given by the vector $\underline{\lambda}_k$ which solves

$$\min_{\underline{\lambda}_k} \{ \| \nabla f(\underline{x}_k) - N_m \underline{\lambda}_k \|^2 \} \tag{1.1.58}$$

The minimising vector is given by

$$\underline{\lambda}_k = (N_m^T N_m)^{-1} N_m^T \nabla f(\underline{x}_k) \tag{1.1.59}$$

The solution of (1.1.38) may be used to obtain an estimate of the Lagrange multipliers at $\underline{x}_k + \underline{d}_1^*$. This may be obtained by considering that at the solution of (1.1.38) there must exist Lagrange multipliers, $\underline{\lambda}_q$, such that

$$N_m \underline{\lambda}_q = \nabla q(\underline{d}_1^*) \tag{1.1.60}$$

holds, whence

$$\underline{\lambda}_q = (N_m^T N_m)^{-1} N_m^T (G_k \underline{d}_1^* + \nabla f(\underline{x}_k)) , \tag{1.1.61}$$

and by (1.1.28) $\underline{\lambda}_q$ is an approximation to

$$(N_m^T N_m)^{-1} N_m^T \nabla f(\underline{x}_k + \underline{d}_1^*) .$$

Let the Lagrange multiplier of (1.1.44) be denoted by $\underline{\lambda}$. Writing the optimality condition of (1.1.44) as

$$H_k \nabla f(\underline{x}_k) + \underline{d}_1^* = H_k N_m \underline{\lambda}$$

and using (1.1.43), (1.1.53), it can be verified that $N_m\underline{\lambda} = -N_m\underline{\lambda}^*$. From assumption (1.2), it follows that N_m has full column rank, thus $\underline{\lambda} = -\underline{\lambda}^*$. Furthermore, using the above optimality condition with (1.1.60) yields

$$N_m \underline{\lambda} = \nabla f(\underline{x}_k) + G_k \underline{d}_1^* = N_m \underline{\lambda}_q . \tag{1.1.62}$$

Again, since N_m has full column rank, it follows that

$$\underline{\lambda} = \underline{\lambda}_q = -\underline{\lambda}^* .$$

If the vector $\underline{d}_c$ is defined as

$$\underline{d}_c = \underline{x}_c - \underline{x}_k \quad (1.1.63)$$

the multiplier $\underline{\lambda}_k$ given by (1.1.59) is a first order approximation to $\underline{\lambda}_c$ since

$$\| \underline{\lambda}_c - \underline{\lambda}_k \| = \| (N_m^T N_m)^{-1} N_m^T (\nabla f(\underline{x}_c) - \nabla f(\underline{x}_k)) \| = O(\| \underline{d}_c \|) \quad (1.1.64)$$

However, $\underline{\lambda}$ and $\underline{\lambda}_q$ are second order approximations due to the relationship

$$\| \underline{\lambda}_c - \underline{\lambda}_q \| = \| (N_m^T N_m)^{-1} N_m^T (\nabla f(\underline{x}_c) - \nabla f(\underline{x}_k) - G_k \underline{d}_1^*) \|$$

$$= O(\| \underline{d}_c \|^2) \quad (1.1.65)$$

In terms of numerical computation, it is clear that when the number of active constraints at $\underline{x}_k$ is high the dimension of the $(n-m) \times (n-m)$ matrix $(Z^T G_k Z)$ in (1.1.49) becomes smaller. On the other hand the dimension of the $m \times m$ matrix $(N_m^T H_k N_m)$ in (1.1.54) gets smaller as the number of active constraints decreases. For the numerical computation of (1.1.54), the $L D L^T$ Cholesky decomposition of G_k and the LQ decomposition of N_m may be used (see, Gill and Murray (1974,b) and Goldfarb (1975)).

Projection algorithms for nonlinear constrained optimisation problems generate their descent directions either by solving a variant of the problem (1.1.47) or a variant of (1.1.51). However, most of them do not use the actual value of the current Hessian matrix, G_k, but prefer to use updating formulae to compute a positive definite approximation to G_k at every $\underline{x}_k$. Sometimes approximations are made to the inverse of G_k or to operators that involve H_k or to the inverse Hessian of the objective function in a given subspace $\Omega \subset E^n$.

1.1.3 Methods Based on Computing Bases for $\bar{\Omega}$ [1]

In this section methods using approximations of G_k or $H_k = G_k^{-1}$ or operators that involve H_k will be discussed. These methods are based on solutions to problem (1.1.51) in Section (1.1.2) and thus use (1.1.50) to compute $\underline{d}_1^*$.

The first algorithm that applied projections to constrained optimisation was Rosen's Gradient Projection method (Rosen (1960)). The descent directions and the Lagrange multipliers generated by this algorithm may be obtained by setting $G_k = I$ in (1.1.51). The directions generated by this algorithm are basically the steepest descent directions projected into the intersection of the active constraints. One method which updates H_k is due to Murtagh and Sargent (1969). They use the rank-one updating formula, (4.1.22) with (4.1.24), for updating an approximation to H . A method which updates an approximation to the operator

$$P_k \triangleq P[H_k] \triangleq (I - H_k N_m (N_m^T H_k N_m)^{-1} N_m^T) H_k \qquad (1.1.66)$$

using the Davidon -Fletcher-Powell (DFP) updating formula ((4.1.22) with $\beta_k = 0$) is due to Goldfarb (1969). All three of the above

---

1 The algorithms discussed in this and the following sections may also be classified according to the subspace in which they approximate G_k. All the algorithms in this section using variable metric formulae, with the exception of Murtagh and Sargent's algorithm and Fletcher's (1972) Method of Hypercubes, generate approximations to G_k, its inverse, or an operator involving the inverse in the subspace $\bar{\Omega}$. The Method of Hypercubes and Murtagh and Sargent's algorithm generate approximations in E^n.

methods have been extended for solving linear inequality constrained problems. They all employ strategies which minimise the objective function on one face of the constraint polytope. This face is changed only when the search for a minimum along the projected search direction $\underline{d}_1^*$ (1.1.54) encounters another constraint (which is not one of those constraints in the intersection of which $\underline{d}_1^*$ lies) or when the minimum value of the objective function on this face has been attained and the objective function may be decreased further only by moving off this face. Thus successive points generated by these methods lie on the same face of the constraint polyhedron. One exception to this is Murtagh and Sargent's method which allows univariate search along $\underline{d}$ in (1.1.43) until a constraint is transgressed whereupon this algorithm also adopts a strategy similar to the above algorithms.

Davidon (1959) extended his algorithms for unconstrained minimisation to solve the linear equality problem by suggesting that the initial approximation to the inverse Hessian is chosen to be in the null space of the matrix N_m^T. One such matrix is the orthogonal projection operator

$$P_o = P[I] = I - N_m(N_m^T N_m)^{-1} N_m^T$$

which projects all vectors in E^n onto the subspace Ω_o with respect to the Euclidean norm (i.e. (1.1.42) with $G_k=I$). Thus it can be established by inspection that

$$N_m^T P_o = 0.$$

This property ensures that the direction of search computed using the approximation P_o,

$$\underline{d}_1^* = P_o \nabla f(\underline{x}_o)$$

satisfies $N_m^T \underline{d}_1^* = \underline{o}$. Hence $\underline{d}_1^*$ is feasible (i.e. $\underline{d}_1^* \in \Omega_o$). The inverse Hessian approximation is updated with incoming curvature information, obtained along $\underline{d}_1^*$. As above, to maintain the feasibility of subsequent search directions, these approximations always remain in the null space of N_m^T . The updating is done using the DFP formula however, the rank-one (see Section (4.1.2)) or the BFGS (Broyden-Fletcher-Goldfarb-Shanno, given by (4.1.22) with (4.1.25)) formulae may be used equally well. The updating of P using a general variable metric formula, of which DFP, BFGS and rank-one are three special cases, is discussed by Powell (1974,a).

Goldfarb (1969) applied the techniques developed by Rosen (1960) to extend Davidon's method to inequality constraints. For the equality constrained case with a quadratic function, Goldfarb proved convergence in n-m iterations and the approximation to P_k updated by the algorithm becomes

$$P_{n-m} = P[H] = H - H N_m (N_m^T H N_m)^{-1} N_m^T H$$

where H is the true inverse Hessian of the quadratic function. Only the first order approximations to the Lagrange multipliers

$$\underline{\lambda}_k = (N_m^T N_m)^{-1} N_m^T \nabla f(\underline{x}_k)$$

are computed in Goldfarb's method since the approximation to P_k always remains in the null space of N_m^T . A disadvantage of Goldfarb's method, however is that if a constraint is dropped from the active set such that the rank of the approximation to P_k is increased by one, no information about the curvature of the objective function is available in the direction of the normal of this constraint.

Murtagh and Sargent (1969) have attempted to overcome this by updating directly the approximation to the inverse Hessian rather than (1.1.66). The projection operator $P[\hat{H}_k]$ is then constructed given $\hat{H}_k$, the approximation to H_k. In this method, the rank-one formula is used to update $\hat{H}_k$. As this formula does not require exact univariate minimisation it is preferable, in this respect, to other variable metric formulae (e.g. BFGS, DFP) which impose such a condition. The search for a minimum along a line may hit an inequality constraint not in the current active set before reaching the minimum. This constraint has to be added to the active set and updating the matrix $\hat{H}_k$ at this point may be done by means of the rank-one formula. Gill and Murray (1974,b) argue that updating $\hat{H}_k$ in E^n rarely provides curvature information in the direction of the normals of the active constraints. Since the descent directions, $\underline{d}_1^*$, have to be orthogonal to these normals to be feasible, the updates of $\hat{H}_k$ contain curvature information in these directions only. Exceptionally, when movement off a constraint occurs near the point of its addition to the active set such curvature information becomes significant. If a constraint is dropped long after it has become active, the curvature information in the direction of its normal, obtained prior to its becoming active,has no longer any significance. Also since $N_m^T \underline{d}_1^* = \underline{o}$, if the BFGS formula is used to update $\hat{H}_k$ for descent steps taken in the intersection of the same constraints, then, by inspection of the BFGS formula, it follows that

$$(N_m^T \hat{H}_{k+1} N_m) = (N_m^T \hat{H}_k N_m).$$

Thus the Lagrange multipliers computed according to

$$\underline{\lambda}_k = (N_m^T \hat{H}_k N_m)^{-1} N_m^T \hat{H}_k \nabla f(\underline{x}_k)$$

may no longer be regarded as second order estimates and similarly the descent direction

$$\underline{d}_1^* = - P[\hat{H}_k] \nabla f(\underline{x}_k)$$

is not equivalent to (1.1.54).

Under certain assumptions on the updating formulae, the equivalence of updating approximations to P_k and updating $\hat{H}_k$ then computing $P[\hat{H}_k]$ is established by Powell (1974,a) (see Section (4.1.2)). Powell (1974,a) has also proved the convergence of Goldfarb's algorithm for a wider class of updating formulae and for inequality constraints. The rate of convergence for quadratic functions and under the assumption of perfect linear searches is established to be $n+\ell$ where ℓ is the number of faces of the constraint polytope over which the search for an optimum is made.

The rest of the methods discussed in this section are those which solve efficiently a series of linear or quadratic programming subproblems. These algorithms may be seen as extensions of projection methods where the restriction $\underline{d}_1 \in \Omega_0$ is relaxed to $\underline{d}_1 \in R$ and $\underline{d}_2$ is effectively determined by the normals of the constraints active at the solution of (1.1.68) below.

The Method of Hypercubes, due to Fletcher (1972), generates quadratic approximations

$$\hat{q}(\underline{x}_{k+1}) = f(\underline{x}_k) + \langle \nabla f(\underline{x}_k), \underline{x}_{k+1} - \underline{x}_k \rangle + \tfrac{1}{2} \langle \underline{x}_{k+1} - \underline{x}_k, \hat{G}_k(\underline{x}_{k+1} - \underline{x}_k) \rangle \tag{1.1.67}$$

to $f(\underline{x}_{k+1})$. To make a realistic approximation, the step size is restricted. Thus, the method involves successive solutions to the problem

$$\underline{x}_{k+1} = \arg\min \{\hat{q}(\underline{x}) \mid N_m^T \underline{x} \geq \underline{b}_{m_o} , \| \underline{x} - \underline{x}_k \|_\infty \leq \delta_\infty\}. \quad (1.1.68)$$

Assuming (1.1.67) is a positive definite approximation, its unconstrained optimum is given by

$$\nabla \hat{q}(\underline{x}_k) = f(\underline{x}_k) + \hat{G}_k(\underline{x}_k - \underline{x}_u) = \underline{o}$$

$$\underline{x}_u = \underline{x}_k - \hat{G}_k^{-1} f(\underline{x}_k) . \quad (1.1.69)$$

By Theorem (3.1) in Section (3.2.3) the solution of (1.1.68) may be expressed as the projection problem

$$x_{k+1} = \arg\min\{\tfrac{1}{2} \| \underline{x} - \underline{x}_u \|^2_{\hat{G}_k} \mid N_{m_o}^T \underline{x} \geq \underline{b}_{m_o} , \| \underline{x} - \underline{x}_u - \hat{H}_k \nabla f(\underline{x}_k) \|_\infty \leq \delta_\infty \} . \quad (1.1.70)$$

Alternatively, (1.1.70) can be reformulated using $\underline{d}_1 = \underline{x} - \underline{x}_k$, $\underline{d} = -\hat{H}_k \nabla f(\underline{x}_k)$ and $\underline{d}_1^* = \underline{x}_{k+1} - \underline{x}_k$. The matrix $\hat{G}_k$ is updated using a rank 2 formula. An earlier variant of (1.1.68), without the constraint on the step size, was developed by Wilson (1963). In this approach, G_k is evaluated at every iteration. Wilson's algorithm is described by Beale (1967) and Wilde and Beightler (1967).

The final method that could be classified in this section is the method of feasible directions which involves the solution of a sequence of problems of the type

$$\max \{- < \nabla f(\underline{x}_k) , \underline{x} - \underline{x}_k > \mid N_{m_o}^T (\underline{x} - \underline{x}_k) \geq \underline{o} , \| \underline{x} - \underline{x}_k \|^2 \leq 1\}$$

where $\underline{x}_k$ is a feasible point. The following result may be used to demonstrate the relationship of this problem with projection methods.

Lemma (1.1)

If $(\underline{x}^* - \underline{x}_k)$ solves the above maximisation problem and $\langle \nabla f(\underline{x}_k), \underline{x}^* - \underline{x}_k \rangle < 0$ then for $\tau > 0$, $\tau(\underline{x}^* - \underline{x}_k)$ also solves

$$\min \{ \|\underline{x} - \underline{x}_k\|^2 \mid N_{m_0}^T(\underline{x} - \underline{x}_k) \geq \underline{0}, -\langle \nabla f(\underline{x}_k), \underline{x} - \underline{x}_k \rangle = 1 \}$$

The proof of this well known result is given by Dennis (1959), Zoutendijk (1960) and Lemke(1961). □

Using the constraint $-\langle f(\underline{x}_k), \underline{x} - \underline{x}_k \rangle = 1$, the objective function $\|\underline{x} - \underline{x}_k\|^2$ may be written as

$$\begin{aligned} \|\underline{x} - \underline{x}_k\|^2 &= \|\underline{x} - \underline{x}_k\|^2 + 2\langle \nabla f(\underline{x}_k), \underline{x} - \underline{x}_k \rangle + 2 \\ &\quad + \|\nabla f(\underline{x}_k)\|^2 - \|\nabla f(\underline{x}_k)\|^2 \\ &= \|\underline{x} - \underline{x}_k + \nabla f(\underline{x}_k)\|^2 - \|\nabla f(\underline{x}_k)\|^2 + 2. \end{aligned}$$

The constant term $2 - \|\nabla f(\underline{x}_k)\|^2$ can be ignored as it does not affect the position of $\underline{x}$ solving the optimisation. Thus, for $N_{m_0}^T \underline{x}_k \geq \underline{b}_{m_0}$, the quadratic optimisation problem in Lemma (1.1) computes the projection of $\underline{x}_k - \nabla f(\underline{x}_k)$ onto

$$\{ \underline{x} \in E^n \mid N_{m_0}^T \underline{x} \geq \underline{b}_{m_0}, \; -\langle \nabla f(\underline{x}_k), \underline{x} - \underline{x}_k \rangle = 1 \}.$$

The second order feasible direction methods described by Polak (1971) may be interpreted as projection methods since they solve quadratic programming subproblems. Second order algorithms for minimising quadratic objective functions subject to nonlinear constraints are discussed in Section (3.4.3). Second order algorithms for nonlinear programming problems are discussed in Section (1.2.1) and Chapter 4.

1.1.4 Methods Based on Computing Bases for Ω_0

In (1.1.34), Ω_0 was defined as the n-m dimensional subspace of E^n. In this section methods with descent directions based on the solution to the unconstrained optimisation problem (1.1.49) in Ω_0 will be discussed. The motivation behind the algorithms in this section is the unconstrained minimisation of the function projected onto the n-m dimensional subspace defined by the active constraints. A major problem with these methods is the determination of a basis of the subspace Ω_0. Let the n x n matrix T be defined as

$$T \triangleq \begin{bmatrix} N_m^T \\ \hline V^T \end{bmatrix} \tag{1.1.71}$$

where the columns of the n x m matrix N_m are the normals to the linearly independent active constraints and V is an n x (n-m) matrix chosen such that T is non-singular. The following result is useful in determining a basis for Ω_0.

Theorem (1.1)

The last n-m columns of the matrix T^{-1} span the subspace Ω_0.

Proof

Since

$$\begin{bmatrix} N_m^T \\ --- \\ V^T \end{bmatrix} T^{-1} = I \qquad (1.1.72)$$

by definition (1.1.71), post multiplying (1.1.72) by $\underline{e}_i$, the ith column of the identity matrix I, yields

$$\begin{bmatrix} N_m^T \\ --- \\ V^T \end{bmatrix} \underline{t}_i^{-1} = \underline{e}_i \qquad (1.1.73)$$

where $\underline{t}_i^{-1}$ denotes the ith column of T^{-1}. Let t_j denote the jth row of T, then (1.1.73) implies that

$$< \underline{t}_i^{-1}, \underline{t}_j > = o , \quad i \neq j .$$

Thus

$$< \underline{t}_j , \underline{t}_i^{-1} > = o \qquad j = 1, \ldots, m ; \; i = m+1, \ldots, n$$

and since by definition

$$N_m \triangleq [\underline{t}_1, \ldots, \underline{t}_m] ,$$

the relationship

$$N_m^T \underline{t}_i^{-1} = \underline{o} \qquad i = m+1, \ldots, n$$

holds. □

According to (1.1.34) in Definition (1.1) this establishes that the n-m vectors $\underline{t}_i^{-1}$ form a basis for the subspace Ω_o. The columns of the n x (n-m) matrix Z defined in (1.1.45) may therefore be set equal to the vectors $\underline{t}_i^{-1}$, i = m+1, ..., n, i.e.

$$Z = \left[\underline{t}^{-1}_{m+1} , \ldots , \underline{t}^{-1}_{n} \right] . \qquad (1.1.74)$$

Second order estimates to the Lagrange Multipliers may be obtained using (1.1.60)

$$N_m \underline{\lambda}_q = \left[N_m \mid V \right] \begin{bmatrix} \underline{\lambda}_q \\ \hline \underline{o} \end{bmatrix} = T^T \begin{bmatrix} \underline{\lambda}_q \\ \hline \underline{o} \end{bmatrix} = \nabla q(\underline{d}_1^*) = G\, \underline{d}_1^* + \nabla f(\underline{x}_k) .$$

(1.1.75)

To invert T^T the identity

$$(T^T)^{-1} = \left[N_m \mid V \right]^{-1} = \begin{bmatrix} (N_m^T N_m)^{-1} N_m^T \\ \hline Z^T \end{bmatrix}$$

is used. Thus

$$\begin{bmatrix} \underline{\lambda}_q \\ \hline \underline{o} \end{bmatrix} = \begin{bmatrix} (N_m^T N_m)^{-1} N_m^T \\ \hline Z^T \end{bmatrix} [G\, \underline{d}_1^* + \nabla f(\underline{x}_k)] \qquad (1.1.76)$$

is obtained. The first order estimates may be obtained by ignoring the second order term in (1.1.76). It should be noted that $\underline{d}_1^*$ is given by (1.1.49).

The first method to use T^{-1} to generate the matrix Z was the reduced-gradient method of Wolfe (1967). In this method the columns of V^T are selected from the normals of inactive constraints. The descent direction of this method is effectively $\underline{d}_1 = -Z\, Z^T \nabla f(\underline{x}_k)$ (see Gill and Murray (1974,a)) and is not a direction given by (1.1.49) even when $G_k = I$. An alternative choice for V is provided by the variable reduction method due to McCormick (1970,a). In the variable reduction method the columns of V^T are chosen from the columns of the identity matrix I such that T is given by

$$T = \left[\begin{array}{c|c} (N^1)^T & (N^2)^T \\ \hline 0 & I \end{array}\right] \quad \text{yielding} \quad Z = \left[\begin{array}{c} -(N^1)^{-T} (N^2)^T \\ \hline I \end{array}\right]$$

(1.1.77)

where N^1 and N^2 are respectively $m \times m$ and $(n-m) \times m$ sub-matrices of N_m such that

$$N_m^T = \left[(N^1)^T \,\middle|\, (N^2)^T \right] . \qquad (1.1.78)$$

The matrix $(N^1)^T$ corresponds to the first m elements of the vector $\underline{x} \in E^n$ and $(N^2)^T$ to the remaining (n-m) elements. The DFP formula is used with a resetting feature for the approximation to H_k in Ω_o. A further extension to this algorithm has been made by McCormick (1970,b) by allowing the computation of G_k in E^n and then computing $(Z^T G_k Z)$ used in (1.1.49).

With (1.1.77) the variable reduction method involves the definition of dependent and independent variables. Accordingly, the first m elements of $\underline{x}$ are set as the vector of dependent variables, $\underline{x}_1$, and the last (n-m) elements as, x_2, the vector of independent variables. Thus the constraints

$$N_m^T \underline{x} = \underline{b}_m$$

may be written using (1.1.78) as

$$(N^1)^T \underline{x}_1 + (N^2)^T \underline{x}_2 = \underline{b}_m$$

and since N^1 is a square matrix, chosen to be non-singular,

$$\underline{x}_1 = (N^1)^{-T} \underline{b}_m - (N^1)^{-T} (N^2)^T \underline{x}_2$$

is substituted into the objective function

$$f(\underline{x}) = f(\underline{x}_1, \underline{x}_2) = f((N^1)^{-T} (N^2)^T \underline{x}_2, \underline{x}_2) .$$

Hence the linear equality constrained problem (1.1.39) may be reduced to an unconstrained optimisation problem in $\underline{x}_2$.

Another method which uses this idea of minimising $f(\underline{x}_2)$ is due to Ganzhella (1970). In this method a variable metric method is used to minimise the objective function in the linear manifold Ω (1.1.35). The choice for V in this method is, in effect, the same as in the variable reduction method.

Gill and Murray (1974,a) have pointed out that a poor choice of V may lead to an ill conditioned T and thus to an unduly large Z. This affects $Z^T\nabla f(\underline{x}_k)$ and (Z^TG_kZ). The former will be large even when $\underline{x}_k$ is near a stationary point and could result in poor estimates of the Lagrange multipliers. It also might cause the Lagrange multipliers to be computed at the wrong time if the criterion to compute them is based on $||Z^T\nabla f(\underline{x}_k)||$. The latter might have an adverse effect on the computation of the direction of search d_1^* (1.1.49). This is due to the ill-conditioning of (Z^TG_kZ) which is reflected in its condition number $K(Z^TG_kZ)$. For a positive definite G_k, this condition number is bounded by the inequality

$$K(Z^TG_kZ) \le K(G_k) [K(Z)]^2$$

(see, Gill and Murray (1974,a)). The choice for Z recommended by Gill and Murray makes this number dependent only on the conditioning of G_k. This entails an LQ factiorisation of N_m^T such that L is an m x m lower triangular matrix and Q an orthogonal matrix (i.e. $Q Q^T = Q^T Q = I$),

$$N_m^T = \left[L \,\vdots\, 0 \right] Q \,. \tag{1.1.79}$$

Furthermore, Q is partitioned such that its first m rows are set equal to the $m \times n$ submatrix Q_1 and the last $n-m$ rows are denoted as Q_2, i.e.,

$$Q = \left[\begin{array}{c} Q_1 \\ \hline Q_2 \end{array} \right] . \tag{1.1.80}$$

Note that

$$N_m^T \left[\begin{array}{c} Q_1 \\ \hline Q_2 \end{array} \right]^T = \left[L \,\vdots\, 0 \right]$$

thus

$$N_m^T Q_2^T = 0.$$

Hence if Q_2 is used as V^T then

$$T = \left[\begin{array}{c} N_m^T \\ \hline Q_2 \end{array} \right]$$

and

$$(T^T)^{-1} = \left[\begin{array}{c} (N_m^T N_m)^{-1} N_m \\ \hline Q_2 \end{array} \right] .$$

Thus the matrix Z is chosen to be Q_2^T and such a choice results in a minimum condition number and for a positive definite G_k

$$K\,(Z^T G_k Z) \le K\,(G_k) \,.$$

Consequently, the conditioning of the algorithms now depend on that of the problem only. Gill and Murray (1973) use this choice of Z for the computation of $\underline{d}_1^*$ (1.1.49) and update their approximation to $(Z^T G_k Z)$ using the rank 2 BFGS formula . An alternative to this method is given by Gill and Murray (1972) where the Hessian G_k is computed. Both these methods are also implementations of the idea of minimising $f(\underline{x}_2)$. Buckley (1975) also discusses a variant of Goldfarb's algorithm that uses the transformation T with the columns of V chosen from the normals of the inactive constraints such that T is full rank.

The computation of the Cholesky factors of G_k, $Z^T G_k Z$, or operators involving H_k, the orthogonal factorisations of N and the modification of these factorisations at each iteration have been explored by Bartels et. al. (1970), Gill et. al. (1974) and Goldfarb (1975).

To end this section, it should be noted that reduced gradient methods may be generalised in such a way that projection methods may be expressed as a special case. This is discussed by Sargent (1974). Since the emphasis in this chapter is the explanation of some constrained optimisation methods in terms of projections, this approach was not adopted.

1.2 EXTENSIONS OF PROJECTION ALGORITHMS FOR SOLVING THE LINEAR EQUALITY CONSTRAINED PROBLEM

In Section (1.1) algorithms for solving the linear equality constrained nonlinear programming problems were discussed. The different approaches to extend these algorithms to solve nonlinear constrained problems and problems with inequality constraints will be discussed in this section.

1.2.1 Extension for Nonlinear Constraints

In Section (1.1.1) it was noted that the Kuhn-Tucker optimality conditions refer to a global optimum when the objective function is convex, the set R (1.1.22) is convex so that each inequality $g_i(\underline{x})$, i=1, ..., m_0 forming a boundary of R is concave. The algorithms discussed below generally do not assume the concavity of $g_i(\underline{x})$ and therefore only their convergence to local optima can be demonstrated.

The extension of the initial projected gradient method for linear constraints to solve nonlinear constrained problems was developed by Rosen (1961). This basically consists of generating first order expansions of the nonlinear constraints about $\underline{x}_k$ and solving a linearly constrained problem similar to (1.1.44) with $G_k = I$. Implicit to this formulation is the solution $\underline{d}_1^*$, to the problem

$$\min \{ \| -\nabla f(\underline{x}_k) - \underline{d}_1 \|^2 \mid N_m^T(\underline{x}_k)\, \underline{d}_1 = \underline{o} \} \qquad (1.2.1)$$

where

$$N_m(\underline{x}_k) \triangleq [\nabla g_1(\underline{x}_k), \ldots, \nabla g_m(\underline{x}_k)] \qquad (1.2.2)$$

and $\underline{x}_k$ is a feasible point within a specified tolerance. The points along $\underline{d}_1^*$ are infeasible with respect to the original nonlinear constraints and a correction procedure is given to obtain a feasible point at which the value of the objective function is lower than $f(\underline{x}_k)$. The point generated along the descent direction $\underline{d}_1^*$ in the intersection of the linear approximations to the active constraints about $\underline{x}_k$,

$$\underline{x}_j = \underline{x}_k + \alpha_k \, \underline{d}_1^* / \| \underline{d}_1^* \| \quad ; \quad j = o \tag{1.2.3}$$

is thus corrected by

$$\underline{x}_{j+1} = \underline{x}_j - N_m(\underline{x}_k) \left(N_m^T(\underline{x}_k) N_m(\underline{x}_k) \right)^{-1} \underline{g}(\underline{x}_j) \tag{1.2.4}$$

for $j=0, 1, \ldots, \nu$ until $\underline{x}_j$ satisfies the nonlinear constraints $g(\underline{x}_\nu) \geq \underline{o}$ to within a specified tolerance, whence $\underline{x}_{k+1} = \underline{x}_\nu$. The convergence of this method is slow. One reason for this is that it uses the steepest descent direction. Polak (1971) discusses modifications to the original gradient projection algorithm for linear constraints to accelerate its convergence. Abadie and Carpentier's (1969) generalisation of Wolfe's (1967) reduced gradient method for nonlinear constraints is also based on steps of descent and correction. Each descent step is taken along the reduced gradient direction and is then corrected to satisfy the nonlinear constraints by successive linearisation of these constraints.

Another extension of projection methods in order to solve non-linear constraints is separating the linear and nonlinear constraints of the problem and redefining the objective function to include a penalty term prohibiting the violation of the nonlinear constraints. Projection methods may then be used to minimise the new function subject to the linear constraints of the problem. This approach is discussed by Fiacco and McCormick with an example as to how Rosen's gradient projection method may be used to solve this problem. They have also reported (Zoutendijk (1966)) that their implementation of this approach has reduced the computer time of a 100 variable problem from 13 to 6 minutes by just treating the nonnegativity constraints separately as linear constraints. Goldfarb (1969) also recommends this approach in conjunction with his conjugate gradient algorithm. It is certain that projection algorithms, because of their simple way of treating linear constraints, would be an asset to methods which set up penalty terms in the objective function to stop the violation of the nonlinear constraints. A more recent method combining penalty functions and gradient projection is described by Luenberger (1974). This procedure involves a new correction step and its convergence rate is established to be independent of the magnitude of the penalty term.

Using the first order expansions about $\underline{x}_k$ of the nonlinear constraints

$$\underline{g}^k(\underline{x}) \triangleq \underline{g}(\underline{x}_k) + N_{m_o}^T(\underline{x}_k)(\underline{x} - \underline{x}_k) \tag{1.2.5}$$

with

$$N_{m_o}(\underline{x}_k) \triangleq \nabla\underline{g}(\underline{x}) \triangleq [\nabla g_1(\underline{x}_k), \nabla g_2(\underline{x}_k), \ldots, \nabla g_{m_o}(\underline{x}_k)] \tag{1.2.6}$$

in conjunction with Lagrangians is discussed by Rosen and Kreuser (1972) and Robinson (1972). The approach adopted by the former is the solution of a sequence of problems of the type

$$\min \{ f(\underline{x}) - \sum_{\substack{j\in I(\underline{x}_k)+V(\underline{x}_k) \\ \lambda_j>o}} \lambda_j(\underline{x}_k)\, g_j(x) \mid \underline{g}^k(\underline{x}) \geq o \} . \tag{1.2.7}$$

The objective function is essentially a Lagrangian penalising the violation of those nonlinear constraints which were violated or active at $\underline{x}_k$. $I(\underline{x}_k)$ is the index set of active nonlinear constraints at $\underline{x}_k$ defined as

$$I(\underline{x}_k) \triangleq \{ i \mid g_i(\underline{x}_k) = o \} \tag{1.2.8}$$

and $V(\underline{x}_k)$ is the index set of violated nonlinear constraints at $\underline{x}_k$

$$V(\underline{x}_k) \triangleq \{ v \mid g_v(\underline{x}_k) < 0 \} . \qquad (1.2.9)$$

The multiplier $\underline{\lambda}(\underline{x}_k)$ is computed at every solution, $\underline{x}_k$, of (1.2.7) according to

$$\underline{\lambda}(\underline{x}_k) = -(N_m^T N_m)^{-1} N_m^T \nabla f(\underline{x}_k) \qquad (1.2.10)$$

where the columns of N_m are those of N_{m_o} corresponding to constraints active at $\underline{x}_k$. Thus $\underline{x}_k$ is set to the current solution of (1.2.7) until such a solution satisfies the Kuhn-Tucker conditions (1.1.24 - 1.1.27) of the original problem (1.1.23). In this algorithm the value of $\underline{\lambda}(\underline{x}_k)$ is dependent on the value of the previous solution, $\underline{x}_k$, to (1.2.7). The second algorithm of this type is due to Robinson (1972) and formulates a different objective function which penalises the deviation of the linearised $\underline{g}^k(\underline{x})$ from the actual nonlinear constraints $\underline{g}(\underline{x})$ by solving

$$\min \{ f(\underline{x}) - \langle \underline{\lambda}(\underline{x}_k), \ \underline{g}(\underline{x}) - \underline{g}^k(\underline{x}) \rangle \mid \underline{g}^k(\underline{x}) \geq \underline{0} \} \qquad (1.2.11)$$

where all the constraints are included in the new objective function. Similar to Rosen's (1961) algorithm for nonlinear constraints, this is also a method which requires the solution of a sequence of problems of the type (1.2.11) until convergence to a Kuhn-Tucker point is obtained. Robinson (1972) has proved the quadratic rate of convergence of his algorithm.

To avoid solving subproblems like (1.2.7) or (1.2.11), methods have been developed that solve quadratic programming subproblems taking account of the curvature information of the nonlinear constraints. One algorithm that solves quadratic subproblems is due to Sargent and Murtagh (1973). This method computes a second order approximation about $\underline{x}_k$ to both the nonlinear constraints and to the original objective function and hence formulates a"quadratic" approximation to the Lagrangian

$$q^L(\underline{x}) \triangleq f(\underline{x}_k) + <\nabla f(\underline{x}_k), \underline{x} - \underline{x}_k> + \tfrac{1}{2} <\underline{x} - \underline{x}_k , G_k(\underline{x} - \underline{x}_k)>$$
$$+ <\underline{\lambda}_k , \underline{g}(\underline{x}_k)> \quad + \quad < \underline{\lambda}_k , N_m^T(\underline{x}_k)(\underline{x} - x_k)>$$
$$+ \sum_{i=1}^{m} \lambda_k^i <\underline{x} - \underline{x}_k , G_k^i (\underline{x} - \underline{x}_k)> . \qquad (1.2.12)$$

The m-dimensional vector $\underline{\lambda}_k$ is defined with each element corresponding to the associated constraint normal in the appropriate column of $N_m(\underline{x}_k)$. Only those constraints active at $\underline{x}_k$ could be included in (1.2.12). However, when a constraint is added to or dropped from the active set, this introduces a problem of discontinuity in the matrix

$$G_k^L \triangleq (G_k + \sum_{i=1}^{m} \lambda^i G_k^i) , \quad H_k^L = (G_k^L)^{-1} \qquad (1.2.13)$$

especially when it is being updated by a variable metric formula (see Section (4.1.2)). In order to avoid this, slack variables may be introduced in the inequality constraints thereby transforming all to equality constraints. Thus all the constraints are treated as active throughout the algorithm and are included in (1.2.12). The expression (1.2.12) is minimised subject to successive linearisations of the nonlinear constraints generating a point, $\underline{x}_{k+1}$, satisfying all the constraints. The steps of descent and correction to obtain a feasible point with a lower objective function value than the current one at $\underline{x}_k$ are given by

$$\underline{\lambda}_{j+1} = \underline{\lambda}_j + \frac{1}{\alpha_k}(N^T(\underline{x}_j) H_k^L N(\underline{x}_k))^{-1} \underline{g}(\underline{x}_k) \qquad (1.2.14)$$

$$\nabla q_{j+1}^L(\underline{x}_k) = \nabla f(\underline{x}_k) + N^T(\underline{x}_k)\underline{\lambda}_{j+1} \qquad (1.2.15)$$

$$\underline{x}_{j+1} = \underline{x}_k - \alpha_k H_k^L \nabla q_{j+1}^L(\underline{x}_k) \qquad (1.2.16)$$

for j=0,1,2, ... until $\underline{x}_j$ satisfies all the constraints and is then set to $\underline{x}_{k+1}$. These steps of descent and correction are clearly more interlinked than those corresponding to (1.2.3) and (1.2.4) in Rosen's (1961) algorithm. The rank-one updating formula is used to compute approximations to H_k^L . This algorithm is a variant of the conceptual algorithms called second order feasible direction methods (Polak (1971)). The properties of this algorithm will be discussed further in Section (3.6.4) in an attempt to resolve the case with a quadratic objective function, without using slack variables and thereby avoiding the introduction of all the nonlinear constraints in the objective function (1.2.12).

The majority of the methods that solve quadratic programming subproblems formulate the Lagrangian associated with the nonlinear programming problem (1.1.23)

$$L\,(\underline{x}\,,\underline{\lambda}) = f(\underline{x}) - \langle \underline{g}(\underline{x})\,,\underline{\lambda}\rangle\,. \tag{1.2.17}$$

Each iteration starts with an approximation to the solution of (1.1.23), $\underline{x}_k$, and the corresponding Kuhn-Tucker multipliers $\underline{\lambda}_k$. Then the quadratic programming subproblem

$$\min\,\{\,\langle \nabla f(\underline{x}_k),\,\underline{d}\rangle + \tfrac{1}{2}\langle \underline{d}\,,G\,\underline{d}\rangle \mid N_{m_o}^T\,(\underline{x}_k)\,\underline{d} - \underline{g}(\underline{x}_k) \geq \underline{o}\,\} \tag{1.2.18}$$

is solved. The vector $\underline{d}$ is the direction along which the next point, $\underline{x}_{k+1}$, is to be determined. Han (1977,a) defines an exact penalty function as a temporary objective function for univariate minimisation along $\underline{d}$. Clearly, the purpose of the penalty function is to prohibit the computation of an $\underline{x}_{k+1}$ that is too infeasible. For the same reason, Powell (1976, 1977,a) considers an alternative using the Kuhn-Tucker multipliers as penalty parameters. The choice of the matrix G is also varied in different implementations. When

G is set to the second derivative terms, with respect to $\underline{x}$, of the Langrangian (1.2.17), Wilson's (1963) algorithm is obtained in which $\underline{x}_{k+1} = \underline{x}_k + \underline{d}$. Garcia - Polomares and Mangasarian (1974) compute an approximation to the second derivative of the Lagrangian, with respect to $\underline{x}$ and $\underline{\lambda}$, and set G to the submatrix containing the second derivative with respect to $\underline{x}$. Powell (1976, 1977,a) has designed a safeguarded procedure to be used with the BFGS (see (4.2.151)) formula for computing approximations to the Hessian of (1.2.17) with respect to $\underline{x}$. Han (1976) has also set G to variable metric approximations to the Hessian of (1.2.17) with respect to $\underline{x}$. Solving (1.2.18) and setting $\underline{x}_{k+1} = \underline{x}_k + \underline{d}$, Han (1976) has discussed specific variable metric formulae that ensure superlinear convergence. The rates of convergence of these methods have also been studied. Wilson's (1963) algorithm, which is basically the analogue of the Newton algorithm in unconstrained optimisation, is quadratically convergent (see, Robinson (1974)). Assuming a convex feasible region and bounded second derivative approximations, Han (1977,a) has proved the global convergence of his algorithm with his step size strategy. Garcia - Palomares and Mangasarian (1974), Han (1976) and Powell (1977,b) have established the R,Q and R superlinear convergence (see Ortega and Rheinboldt (1970, Chapter 9)) of their respective algorithms with $\underline{x}_{k+1} = \underline{x}_k + \underline{d}$. Algorithms that solve the dual of (1.2.18) have been studied by Han (1977,b).

Algorithms that involve quadratic approximations to the penalty function

$$P(\underline{x}\ , r) = f(\underline{x}) + \frac{1}{r} < \underline{g}(\underline{x}), \underline{g}(\underline{x}) > \qquad (1.2.19)$$

instead of the Lagrangian (1.2.17) have been considered by Murray (1969) and Biggs (1974). The scalar r in (1.2.19) is the penalty parameter and $\underline{g}$ is the vector of violated constraints. Murray (1969) has considered an algorithm which solves a sequence of the quadratic programming problems

$$\min \{ < \underline{d} , \nabla f(\underline{x}_k)> + \tfrac{1}{2} < \underline{d} , \hat{G}_k \underline{d} > |$$

$$| N^T(\underline{x}_k) \underline{d} \geq (1 - r_k r_{k-1}^{-1}) \underline{g}(\underline{x}_k) \} \qquad (1.2.20)$$

where r_k is the penalty parameter at iteration k, $N^T(\underline{x}_k)$ is the Jacobian of the vector of violated constraints at $\underline{x}_k$ and $\hat{G}_k$ is an approximation to the Hessian of $P(\underline{x},r)$. Murray has considered suitable steplength strategies for determining successive points such that $\underline{x}_{k+1} = x_k + \alpha \underline{d}$. Biggs (1974) has introduced a class of algorithms that solve a sequence of equality constrained quadratic programming subproblems to generate the descent directions. The subproblems

$$\min \{ < \underline{d}, \nabla f(\underline{x}_k) > + \tfrac{1}{2} < \underline{d}, \hat{G}_k \underline{d} > | N^T(\underline{x}_k) \underline{d} = \frac{r_k}{2} \hat{\underline{\lambda}}_k - g(\underline{x}_k) \}$$

$$(1.2.21)$$

are derived from the optimality conditions of the truncated Taylor series expansion of (1.2.19), i.e.

$$\nabla P(\underline{x}_k + \underline{d} , r_k) = \nabla f(\underline{x}_k) + \frac{2}{r_k} N(\underline{x}_k) \underline{g}(\underline{x}_k) + \frac{2}{r_k} N(\underline{x}_k) N^T(\underline{x}_k) \underline{d} = \underline{o} .$$

The matrix $\hat{G}_k$ is chosen to be an approximation, at x_k, to the Hessian of $f(\underline{x})$ or to the Hessian of the Lagrangian. The vector $\hat{\underline{\lambda}}_k$ is a local estimate for the Lagrange multipliers and a number of ways of computing $\hat{\underline{\lambda}}_k$ is discussed by Biggs (1974, 1975). Subsequent points are computed using $\underline{x}_{k+1} = \underline{x}_k + \alpha \underline{d}$ where the scalar α is chosen to ensure that the decrease $P(\underline{x}_k , r_k) - P(\underline{x}_{k+1}, r_k)$ is "sufficient" (see, Ortega and Rheinboldt (1970, p. 479)).

Finally, a conceptually important and computationally difficult class of algorithms are those which search for $\underline{x}_{k+1}$ along arcs using the given search direction, thus

$$\underline{x}_{k+1} = P_R(\underline{x}_k - \alpha\nabla f(\underline{x}_k)) , \qquad (1.2.22)$$

where P_R is the projection operator projecting vectors onto the feasible set R. R is assumed to be convex and the direction of search is, as indicated in (1.2.22), generally chosen to be $-\nabla f(\underline{x}_k)$. One way of choosing α is by means of the univariate minimisation

$$\min \{ f(P (\underline{x}_k - \alpha\nabla f(\underline{x}_k)) \mid \alpha \geq o\} . \qquad (1.2.23)$$

McCormick and Tapia (1972) discuss choosing α that solves (1.2.23) . Goldstein (1964), Levitin and Polyak (1966) who have initially proposed the method (1.2.22) have suggested choosing α within a range determined by the bound on the Hessian of $f(\underline{x})$. Bertsekas (1976) has developed a generalised Armijo-type step size rule for choosing α . In Chapter 4 some extensions of (1.2.22) to more general descent directions and stepsize strategies will be considered.

1.2.2 Active Set Strategies

In this section further methods for extending algorithms, designed for equality constrained problems, to inequality constraints will be considered. There are basically two types of active set strategies to account for the inequality constraints at every iteration. The first is the simple extension of the idea of slack variables widely used in linear programming. This is discussed in relation to nonlinear constraints by Sargent and Murtagh (1973) as described in Section (1.2.1). A slack variable is conceptually introduced for every inequality constraint thereby transforming all such constraints to equalities. In this way, all constraints are included in the active set and remain there throughout the algorithm. Hence, the active set does not change. The second type of strategy which aims to include only a subset of the inequality constraints in the active set during an iteration allows for some degree of variation as to which constraints to include in the active set, which to drop from it and when to drop them. Such strategies are especially useful for linear inequality constrained problems. Basically those constraints (together with the equality constraints) satisfied as equalities at the current point, $\underline{x}_k$, are included in the active set. A constraint is added to this set when the search direction from $\underline{x}_k$ hits one which is not already in the active set. In the Sargent and Murtagh (1973) algorithm which does not incorporate univariate minimisation along the search direction an additional condition on the degree to which the constraint is violated has to be satisfied: if the violation is too large then the steplength of the search direction is shortened and the violated constraint is not added to the active set. A constraint is dropped from the active set if the search direction in the intersection of the rest of the active constraints (i.e. not including the constraint to be dropped) is a direction of descent and is feasible with respect to this constraint. As the Lagrange multiplier of a constraint gives the rate of change of the objective function in relation to this constraint (see, e.g., Sargent (1974)), a negative multiplier implies that if the corresponding constraint is dropped then a reduction may be obtained in the objective function value. Thus if a single constraint is to be dropped the one with the most negative multiplier is chosen. Most algorithms do not consider dropping a constraint until a stationary point has been reached in the intersection of the currently active constraints while others involve the antizigzagging rule due to Zoutendijk (1960) to prevent the same constraint from being continuously dropped and the added to the active set. The criteria to drop a constraint are based on the value of the Lagrange multiplier and on tests to stipulate the amount of reduction in the objective function if a constraint is removed. Such criteria may even be formulated in terms of the solution to a complementary pivoting algorithm (Lemke (1968)) as in the algorithm due to Sargent and Murtagh (1973). Comprehensive reviews comparing different tests and strategies to select the active set have been published by Gill and Murray (1974,a,b) and Sargent (1974).

1.3 REVIEW AND ORIGINAL CONTRIBUTIONS

In this chapter a unified approach to the derivation of projection techniques is provided and existing projection methods for nonlinear optimisation are reviewed. The projection techniques discussed in this chapter can be used in various stages of nonlinear programming algorithms. In particular, some of these algorithms require solutions of quadratic minimisation subproblems which can be formulated in terms of quadratic programming. Most quadratic programming algorithms assume that an initial feasible point is known. Such a point can be computed using an algorithm for locating the feasible point of a region constrained with linear inequalities and equalities. The use of projection techniques in these subproblems is discussed in subsequent chapters.

In effect, this work consists of two parts. The first part, Chapters 1 to 4, is concerned with the development of projection techniques for different aspects of constrained optimisation. The computation of a feasible point of a linearly constrained region and the application of projection techniques to this problem are discussed in Chapter 2. A new projection algorithm for computing feasible points is given in Section (2.1.2). The relation of this algorithm with the simplex method of linear programming is exposed in the context of the pivoting rules employed. In Section (2.2.2) a projection algorithm that does not necessarily compute feasible vertices is discussed.

In Chapter 3 a new positive definite quadratic programming algorithm is discussed. This algorithm makes use of the unconstrained optimum of the quadratic objective function (see Theorem (3.1)). The algorithm is especially useful if the unconstrained optimum is known. A statement of the algorithm is given in Section (3.4.1) and its extension to nonlinear constraints is discussed in Section (3.4.3).

In Chapter 4 a nonlinear programming algorithm is discussed. This algorithm, (4.1.7), may be interpreted as a generalisation of the Goldstein-Levitin-Polyak algorithm discussed in Section (1.2.1), (Goldstein (1964)), (Levitin and Polyak (1966)). Algorithm (4.1.7) uses more general projections and descent directions and introduces new stepsize strategies. The convergence properties of the stepsize strategies and the convergence of the algorithm is discussed in Section (4.2). Conditions are derived under which the algorithm achieves unit step lengths. The superlinear rate of convergence of the algorithm is discussed.

The second part of the work is concerned with the application of projection techniques to computational problems in policy optimisation. A fundamental problem in policy optimisation is the choice of a suitable objective function. In Chapter 5 a new approach to this problem is formulated. This is an iterative technique based on rank-one updates to the Hessian of the quadratic objective function. The objective function is thus altered systematically until an optimal solution acceptable to the policy-maker is generated. Projection techniques are used to justify and explore the desirable properties of the method.

Policy optimisation algorithms for large nonlinear econometric models are discussed in Chapter 6. These algorithms are based on extensions of the projection methods in Section (1.1.4) in that, similar to (1.1.77) - (1.1.78), they exploit the natural partition between dependent and independent variables in econometric models. Policy optimisation applications using nonlinear econometric models is a new field in which little computational experience exists. Algorithms for policy optimisation have to be designed to utilise the general characteristics of econometric models. In Section (6.3) a Newton type algorithm and in Section (6.4) its quasi-Newton extension are described. Numerical experience obtained in applications to three different nonlinear econometric models have shown that both algorithms perform satisfactorily even under certain simplifying assumptions.

CHAPTER 2

PROJECTION METHODS FOR COMPUTING FEASIBLE POINTS OF LINEARLY CONSTRAINED REGIONS

2.1 INTRODUCTION

In this Chapter projection concepts are used for computing feasible points of linearly constrained regions. Such feasible points are required by some optimisation algorithms (see Chapter 3). Algorithms developed for this purpose are often associated with linear programming techniques. A projection approach to such algorithms also demonstrates the applicability of the projection techniques discussed in Chapters 1 and 3. Thus, operators used in algorithms for computing feasible points and for quadratic and nonlinear programming can be seen within a unified framework.

In Section (2.1.1.) the basic concepts and the projections of an infeasible point are discussed. In Section (2.1.2) a new projection algorithm is described. Using a dual representation, the relation of this algorithm with the simplex method of linear programming is discussed. The convergence of the algorithm is established by invoking the elegant pivot selection rules due to Bland (1977) (see (2.1.33)). In Section (2.2.1) the degeneracy and redundancy of constraints are discussed. These, along with the results in Section (2.1.1) concerning the linear dependence of constraint normals, establish the geometric interpretation of some of the rules employed by the algorithms. In particular, some of these results help to clarify the relation of the projection algorithm of Section (2.1.2) with the simplex method. A projection algorithm that does not necessarily compute feasible vertices is discussed in Section (2.2.2).

2.1.1 The Feasible Region

Given an infeasible initial point $\underline{x}_u$, the problem is to determine a feasible point $\underline{x} \in R$ where the region R is defined as

$$R \triangleq \{\underline{x} \in E^m \mid N_{m_o}^T \underline{x} \geq \underline{b}_{m_o}\} . \qquad (2.1.1)$$

The subscript m_o denotes the total number of constraints. At an infeasible point $\underline{x}_u \in E^n$ some of the constraints describing R may be violated, others just satisfied as equalities and the rest satisfied as strict inequalities. The set of the indices of all the constraints violated at $\underline{x}_u$ will be defined as

$$V(\underline{x}_u) \triangleq \{v \mid \langle \underline{n}_v, \underline{x}_u \rangle - b_v < o , v \in \{1,2,\ldots,m_o\}\} . \qquad (2.1.2)$$

The vector $\underline{n}_v$ is the normal of the violated constraint and b_v its right hand side. In general, $V(\underline{x})$ will denote the indices violated at $\underline{x}$ and m_v will be the number of constraints in $V(\underline{x})$. Similarly, the set for all the active constraints will be

$$I(\underline{x}_u) \triangleq \{i \mid \langle \underline{n}_i, \underline{x}_u \rangle - b_i = o , i \in \{1,2,\ldots,m_o\}\} \qquad (2.1.3)$$

with $I(\underline{x})$ as the set of active constraints at $\underline{x}$ and m as the number of active constraints as $\underline{x}$. Definitions (2.1.1), (2.1.2) and (2.1.3) correspond to (1.1.22), (1.2.9) and (1.2.8) respectively, stated for linear constraints. Finally for those $m_o - m_v - m$ constraints satisfied at $\underline{x}_u$, the corresponding set of indices is given by

$$S(\underline{x}_u) \triangleq \{s \mid \langle \underline{n}_s, \underline{x}_u \rangle - b_s > 0 \ , \ s \in \{1,2,\ldots,m_0\} \} \qquad (2.1.4)$$

and the set at $\underline{x}$ will be denoted by $S(\underline{x})$. Clearly, one property of the sets S,V,I is that

$$V(\underline{x}) \cap I(\underline{x}) = V(\underline{x}) \cap S(\underline{x}) = I(\underline{x}) \cap S(\underline{x}) = \phi \ , \ \forall \underline{x} \in E^n .$$

A projection approach to the computation of a feasible point on R will be discussed in this section.

Assuming $m_0 \leq n$ and the constraint normals $\underline{n}_i$, $i = 1,\ldots,m_0$ are linearly independent, a feasible point may be obtained by first projecting $\underline{x}_u$ into the intersection of the constraints violated or active at $\underline{x}_u$. The projection may be chosen to be orthogonal in which case

$$\underline{x}_f = \underline{x}_u - N(N^TN)^{-1}(N^T\underline{x}_u - \underline{b}) \qquad (2.1.5)$$

where the columns of N are the normals of the active and violated constraints. Alternatively, the projection might be weighted by a symmetric positive definite matrix, G in which case

$$x_{f'} = \underline{x}_u - G^{-1}N(N^TG^{-1}N)^{-1} (N^T\underline{x}_u - \underline{b}) \ . \qquad (2.1.6)$$

Although $\underline{x}_f$ and $\underline{x}_{f'}$ will not, in general, be the same point, it may be verified directly from (2.1.5) and (2.1.6) that they satisfy the violated and active constraints at $\underline{x}_u$, hence

$$N^T\underline{x}_f - \underline{b} = N^T\underline{x}_u - \underline{b} - N^T N (N^TN)^{-1}(N^T\underline{x}_u - \underline{b}) = \underline{o}$$

and

$$N^T\underline{x}_{f'} - \underline{b} = N^T\underline{x}_u - \underline{b} - N^T G^{-1} N (N^TG^{-1}N)^{-1} (N^T\underline{x}_u - \underline{b}) = \underline{o}.$$

The projection operators used above are the outcome of quadratic programming problems and are discussed in Chapter 3. Furthermore, (2.1.6) involves the same operator as that used in the quadratic programming algorithm in Chapter 3.

If $\underline{x}_f$ or $\underline{x}_{f'}$ violates any further constraints, in $S(\underline{x}_u)$, the normal of the most violated constraint in $S(\underline{x}_u)$, say

$$\langle \underline{n}_v, \underline{x}_{f'} \rangle - b_v = \underset{s \in S(\underline{x}_u)}{\arg\min} \{ \langle \underline{n}_s , \underline{x}_{f'} \rangle - b_s < o \} ,$$

is added as an extra column to N and the projection (2.1.5) or (2.1.6) is recomputed using the new N . This is repeated if the new projection violates any other constraints in $S(\underline{x}_u)$. The process may go on until all $m_o \leq n$ constraint normals are in N . If a feasible point exists, the above procedure would locate one such point, provided the columns of N are linearly independent. At such a point all constraints whose normals are in N are satisfied as equalities and the rest are satisfied as strict inequalities.

The projection operators involved in (2.1.5) and (2.1.6) may be updated when an additional column is added to N . Relevant updating formulae are discussed in Sections (3.5.1) - (3.5.2). When rank (N)=n, the matrix N has n columns and the operators $N(N^TN)^{-1}$ and $G^{-1}N(N^TG^{-1}N)^{-1}$ become $(N^T)^{-1}$ and

$$\underline{x}_f = \underline{x}_{f'} = (N^T)^{-1} \underline{b} . \qquad (2.1.7)$$

Thus, there is a difference between (2.1.5) and (2.1.6) only when rank (N) $\neq$ n . In this case the location of $\underline{x}_{f'}$ is dependent on the chosen weighting matrix G .

When the normals of some constraints describing R (2.1.1) are linearly dependent, the above procedure for computing a feasible point need not work. Assuming that $\underline{x}_f$ is computed using (2.1.5), the normal $\underline{n}$, linearly dependent on the columns of N in (2.1.5), ensures that the corresponding constraint, say

$$<\underline{n} , \underline{x}> - b \geq o ,$$

remains in its satisfied, active or violated state, at $\underline{x}_f$, for any $\underline{x}$ such that

$$N^T (\underline{x}_f - \underline{x}) = \underline{o} .$$

This is because $\underline{n} = N\underline{\lambda}$ where $\underline{\lambda}$ is a nonzero vector describing the linear dependence of $\underline{n}$ on the columns of N and

$$<\underline{n} , \underline{x}_f - \underline{x}> = <N\underline{\lambda} , \underline{x}_f - \underline{x}> = o . \qquad (2.1.8)$$

Thus, the normals of active or satisfied linearly dependent constraints need not be added to N as extra columns. However, linearly dependent violated constraints may have to be added in order to ensure progress towards a point that will satisfy them. When such an addition is required, the normal of the violated constraints has to be exchanged with a column of N such that the resulting N, after the exchange, has linearly independent columns. The linear independence ensures the nonsingularity of (N^TN) or $(N^TG^{-1}N)$in (2.1.5) and (2.1.6). When a constraint normal is dropped from the columns of N, the procedure above cannot prevent this normal from reappearing again as a column of N. Thus, the same constraint normal may be dropped from and subsequently added to N . The same active constraints may reappear and the method may cycle between a number of points. The above approach does not provide a way of ensuring that the total infeasibility of the violated constraints is being reduced while the same constraints reappear in the columns of N . In Section (2.1.3) a simple extension of this method that ensures convergence, in finite number of steps, to a feasible point, will be discussed.

2.1.2 The Linear Dependence of Constraint Normals

When, during the initial setting up of N with the normals of the constraints in $V(\underline{x}_u) \cup I(\underline{x}_u)$ a normal, say $\underline{n}$, is found to be linearly dependent on the already chosen columns of N, this constraint is ignored. If the projection of $\underline{x}_u$ still violates the same constraint, the vector $\underline{\lambda}_f$ or $\underline{\lambda}_{f'}$ is computed by solving either

$$\min_{\underline{\lambda}_f} \{ \|\underline{n} - N\underline{\lambda}_f\|^2 \} \tag{2.1.9}$$

or

$$\min_{\underline{\lambda}_{f'}} \{ \|\underline{n} - N\underline{\lambda}_{f'}\|^2_{G^{-1}} \} . \tag{2.1.10}$$

The solution of (2.1.9) is

$$\underline{\lambda}_f = (N^T N)^{-1} N^T \underline{n} \tag{2.1.11}$$

whereas the solution of (2.1.10) is given by

$$\underline{\lambda}_{f'} = (N^T G^{-1} N)^{-1} N^T G^{-1} \underline{n} . \tag{2.1.12}$$

If the vector $\underline{\lambda}$ ($\underline{\lambda}_f$ or $\underline{\lambda}_{f'}$) has positive elements, one of these, say i, is chosen and the ith column of N, corresponding to the ith constraint, is replaced by $\underline{n}$. The resulting matrix is denoted by $\hat{N}$ and the corresponding right hand side vector is denoted by $\hat{\underline{b}}$. Hence we have

$$\begin{aligned} N &\triangleq [\underline{n}_1, \underline{n}_2, \ldots, \underline{n}_{i-1}, \underline{n}_i, \underline{n}_{i+1}, \ldots.] \\ \underline{b}^T &\triangleq [b_1, b_2, \ldots, b_{i-1}, b_i, b_{i+1}, \ldots.] \\ \hat{N} &\triangleq [\underline{n}_1, \underline{n}_2, \ldots, \underline{n}_{i-1}, \underline{n}, \underline{n}_{i+1}, \ldots.] \end{aligned} \tag{2.1.13}$$

$$\hat{\underline{b}}^T \triangleq [b_1, b_2, \dots , b_{i-1}, b , b_{i+1}, \dots] .$$

where the columns of $N, \hat{N}$ are linearly independent. We shall now prove that the projection of $\underline{x}_u$ using $\hat{N}, \hat{\underline{b}}$ with either (2.1.5) or (2.1.6) also satisfies the constraint i. This is a special case of the following general result.

Theroem (2.1)

If $$N^T \underline{x}_f = \underline{b} \quad (2.1.14)$$

and $$\langle \underline{n} , \underline{x}_f \rangle < b \quad (2.1.15)$$

and if among the elements of $\underline{\lambda}$ satisfying

$$\underline{n} = N\underline{\lambda} \quad (2.1.16)$$

there is at least one, say λ_i, which is strictly positive, then

$$\{\underline{x} \in E^n \mid \langle \underline{n}_i , \underline{x} \rangle \geq b_i\} \supset \{\underline{x} \in E^n \mid \hat{N}^T \underline{x} = \hat{\underline{b}} \} . \quad (2.1.17)$$

Proof

The proof may be given either by formulating the above statement as a theorem of the alternative (see, Mangasarian (1969)) or by proving the inclusion (2.1.17) directly. The following is a direct proof.

Using (2.1.14), (2.1.15) and (2.1.16)

$$b > \langle \underline{n}, \underline{x}_f \rangle = \langle N\underline{\lambda}, \underline{x}_f \rangle = \langle \underline{\lambda}, \underline{b} \rangle , \quad \lambda_i > 0 , \quad (2.1.18)$$

is obtained. Using (2.1.16) and the definition of $\hat{N}$ (2.1.13) yields

$$\underline{n}_i = \hat{N}\,\hat{\underline{\lambda}} \tag{2.1.19}$$

where $\hat{\lambda}_i > o$ since $\lambda_i > o$. Furthermore,

$$b_i < \langle \hat{\underline{\lambda}}, \hat{\underline{b}} \rangle \tag{2.1.20}$$

and

$$\langle \underline{n}_i, \underline{x} \rangle = \langle \hat{N}\hat{\underline{\lambda}}, \underline{x} \rangle = \langle \hat{\underline{\lambda}}, \hat{\underline{b}} \rangle > b_i \tag{2.1.21}$$

for all $\underline{x}$ satisfying $\hat{N}^T x = \hat{\underline{b}}$. This yields the result (2.1.17) . □

It should be noted that if none of the elements of $\underline{\lambda}$ are strictly positive when a constraint is still violated, the feasible region R is empty. This is a simple consequence of Theorem (2.1), when $b_i > \langle \hat{\underline{\lambda}}, \hat{\underline{b}} \rangle$ and $\langle \underline{n}_i, \underline{x} \rangle - b_i < o$ for all $\underline{x}$ satisfying $\hat{N}^T \underline{x} = \hat{\underline{b}}$, and will be discussed further in Section (2.2.1).

As a generalisation of Theorem (2.1), we shall give the condition which ensures that the constraint with normal $\underline{n}_i$ of N contains the region described by

$$\{\underline{x} \in E^n \mid \hat{N}^T \underline{x} \geq \hat{\underline{b}}\} \tag{2.1.22}$$

where N and $\hat{N}$ are given by (2.1.13).

Theorem (2.2)

If $\underline{x}_f$ satisfies (2.1.14) and (2.1.15) and $\underline{\lambda}$ satisfies (2.1.16) and if $\underline{\lambda}$ has only one strictly positive element, say λ_i , then

$$\{\underline{x} \in E^n \mid \langle \underline{n}_i, \underline{x} \rangle - b_i \geq o\} \supset \{\underline{x} \in E^n \mid \hat{N}^T \underline{x} - \hat{\underline{b}} \geq o\} \tag{2.1.23}$$

Proof

As Theorem (2.1), this theorem may also be formulated as a theorem of the alternative. However, the following is a direct proof of the above statement.

Using (2.1.14) - (2.1.16) yields

$$b > \langle \underline{n}, \underline{x}_f \rangle = \langle N\underline{\lambda}, \underline{x}_f \rangle = \langle \underline{\lambda}, \underline{b} \rangle \quad , \lambda_i > o \,. \tag{2.1.24}$$

Using (2.1.24), the inequality

$$b_i < \langle \hat{\underline{b}}, \hat{\underline{\lambda}} \rangle \tag{2.1.25}$$

is obtained where all the elements of $\hat{\underline{\lambda}}$ are nonnegative. Furthermore, $\hat{\underline{\lambda}}$ satisfies

$$\underline{n}_i = \hat{N} \hat{\underline{\lambda}} \,. \tag{2.1.26}$$

For $\underline{x}$ satisfying

$$\hat{N}^T \underline{x} - \hat{\underline{b}} \geq \underline{o} \tag{2.1.27}$$

we have, using (2.1.25),

$$\langle \underline{n}_i, \underline{x} \rangle = \langle \hat{N}\hat{\underline{\lambda}}, \underline{x} \rangle \geq \langle \hat{\underline{b}}, \hat{\underline{\lambda}} \rangle > b_i \,. \quad \square \tag{2.1.28}$$

A similar statement related to (2.1.23) is that this relationship holds when there exists a vector of nonnegative real numbers $\hat{\underline{\lambda}}$ (with one strictly positive) such that $\underline{n}_i = \hat{N} \hat{\underline{\lambda}}$ and $\underline{b}_i \leq \langle \hat{\underline{b}}, \hat{\underline{\lambda}} \rangle$. This may be proven either as a special result of Caratheodory's Theorem or the half space on the left hand side of (2.1.23) may be shown to be the consequence of the system on the right hand side. In either case the proofs are given by Rockefeller (1972, p. 160 and 198 respectively). Furthermore, a clear implication of

(2.1.23) is that the constraint on the left is redundant.

The linear dependence of a normal vector on the columns of N can be detected while updating the projection operators to incorporate this new normal. This is discussed further in Section (3.5.3).

2.1.3 A Projection Algorithm

When $m_0 > n$ methods based on (2.1.5) or (2.1.6) that add the normals of the violated constraints as additional columns of N need not work since N can have at most n linearly independent columns. When N has n columns and the point $\underline{x}_f$ given by (2.1.7) still violates some constraints, the normals of these violated constraints can only be exchanged with existing columns of N. In this section a simple algorithm is described that performs these exchanges. It is assumed that there exist n linearly independent constraint normals that define a unique intersection. However, this assumption may also be relaxed. The algorithm uses the two rules given below.

Definition (2.1)

A half space is defined as

$$h_i \triangleq \{\underline{x} \in E^n \mid \langle \underline{n}_i , \underline{x} \rangle \geq b_i \} \qquad (2.1.29)$$

and the hyperplane associated with h_i is given by

$$h_i^o \triangleq \{\underline{x} \in E^n \mid \langle \underline{n}_i , \underline{x} \rangle = b_i \} \quad . \qquad (2.1.30)$$

Rule (2.1.31)

Given the n linearly independent constraint normals in N that define the current intersection, $\underline{x}$, and the normal $\underline{n}$ of a violated constraint at $\underline{x}$, compute $\underline{\lambda}$ to satisfy (2.1.16). Define Λ as the set whose elements are the strictly positive elements of $\underline{\lambda}$. Thus,

$$\Lambda \triangleq \{\lambda_k \mid \lambda_k > 0 \ , \underline{n}=\sum_{j=1}^{n} \underline{n}_j \ \lambda_j \ , \ k \in \{1,2,\ldots,n\} \} . \quad (2.1.32)$$

By Theorem (2.1), (2.1.17) holds if $\lambda_i \in \Lambda$ and the ith column of N,$\underline{n}_i$, is exchanged with $\underline{n}$.

Rule (2.1.33)

Given all the constraints describing the feasible region R, index the constraints sequentially to define the set J

$$J \triangleq \{ j \mid < \underline{n}_j,\underline{x} > - b_j \geq 0 \ , \ j = 1,2,\ldots,m_o\} .$$

The index assigned to each constraint remains the same at all times. If there are violated constraints then choose the violated constraint normal with the lowest index $j \in J$. If this lowest index is j^* , then

$$j^* = \min_{j \in V(\underline{x})} \{ j \mid j \in J \} .$$

The normal $\underline{n}_{j*}$ will be exchanged with an existing column of N and the constraint j* will be required to be active. Each column $\underline{n}_i$, $i = 1,2,\ldots,n$, of N has an index assigned to it in J above. Let k(i) denote the index $k \in J$ corresponding to the normal at the ith column of N. Choose the column of N to be replaced by $\underline{n}_{j*}$, to be the normal that corresponds to a positive element of $\underline{\lambda}$ and also to the lowest index in J. Let k* denote the lowest index, thus

$$k^* = \min \{ k(i) \mid k(i) \in J , \lambda_i \in \Lambda \} .$$

Hence the normal $\underline{n}_{j*}$ of the violated constraint h_{j*} replaces the column in N occupied by the normal $\underline{n}_{k*}$ of the active constraint h_{k*} .

Algorithm (2.1.34)

Step 0: Given an initial point $\underline{x}_u$, compute the vector $\underline{\theta}(\underline{x}_u) = N_{m_0}^T \underline{x}_u - \underline{b}_{m_0}$. If all the elements of $\underline{\theta}(\underline{x}_u)$ are nonnegative stop; $\underline{x}_u$ is feasible.
Otherwise, construct the columns of N from the linearly independent normals of the violated constraints. Choose each subsequent normal such that it has the lowest index $j \in J$ in Rule (2.1.33). If there are less than n violated constraints, complete the n columns of N from the linearly independent normals of the active and satisfied constraints. Thus the n columns of N are linearly independent. Construct the vector $\underline{b}$ of the right hand side elements corresponding to the same order of N.

Step 1: Compute $\underline{x}$ from

$$N^T \underline{x} = \underline{b} \qquad (2.1.35)$$

Step 2: If $V(\underline{x}) = \phi$, stop: $\underline{x}$ is feasible. (Clearly, constraints whose normals are columns of N need not be tested for constraint violation at $\underline{x}$.) Otherwise, choose the violated constraint normal to become a column of N according to Rule (2.1.33), i.e. choose the violated constraint corresponding to the lowest index $j \in J$. Let j* be the corresponding normal.

Step 3: Compute $\underline{\lambda}$ from

$$N\underline{\lambda} = \underline{n}_{j*} \qquad (2.1.36)$$

If $\underline{\lambda} \le \underline{o}$ then $R = \phi$ stop: there is no feasible solution (see Theorem (2.4)). Otherwise construct Λ according to Rule (2.1.31) with $\underline{n} = \underline{n}_{j*}$ in (2.1.32). Choose the normal in N to be replaced by $\underline{n}_{j*}$using Rule (2.1.33), i.e. choose the column of N corresponding to (i) a positive element of $\underline{\lambda}$ and (ii) to the lowest in J. Let this index be k^*.

Step 4: Replace $\underline{n}_{k*}$ in N with $\underline{n}_{j*}$ as the new column, similarly, replace b_{k*} in $\underline{b}$, in (2.1.35), with b_{j*} where b_{k*} and b_{j*} are respectively the right hand sides of the constraints $k^*$and j^*. Go to Step 1.

Rosen (1960) has suggested an algorithm based on (2.1.5), for determining a feasible point, without considering its convergence properties. Rosen's algorithm involves the computation of the projection operator $P = I - N(N^TN)^{-1} N^T$ with which $P\underline{n}_v$, $\underline{v} \in V(\underline{x})$, is computed. If $|P\underline{n}_v| = \underline{o}$ and $\underline{\lambda} \le \underline{o}$, where $\underline{\lambda}$ is given by (2.1.11), the algorithm stops as $R=\phi$; if $|P\underline{n}_v| = \underline{o}$ and $\lambda_i > o$, where λ_i is an element of $\underline{\lambda}$, the corresponding normal, $\underline{n}_i$, in N is replaced by $\underline{n}_v$; if $|P\underline{n}_v| > \underline{o}$, $\underline{n}_v$ is added as an extra column to N . This procedure, on its own, cannot ensure finite convergence as the constraint normals removed from N may reappear in N and after a number of exchanges N may have the same columns. There is no precaution in Rosen's method to ensure convergence to a feasible point after a finite number of column exchanges. In order to ensure finite convergence, Algorithm (2.1.34) starts at the intersection of n constraint hyperplanes. Under this condition, Rules (2.1.31) and (2.1.33) ensure the finite convergence of the algorithm.

To demonstrate the relationship of Algorithm (2.1.34) to the simplex method of linear programming, consider the problem

$$\min \{ <\underline{o},\underline{x}> \mid N_{m_o}^T \underline{x} \geq \underline{b}_{m_o} \} \tag{2.1.37}$$

where $\underline{o}$ is the n-dimensional vector with all its elements zero. The dual of (2.1.37) is

$$\max \{ <\underline{b}_{m_o} , \underline{u}> \mid N_{m_o} \underline{u} = \underline{o} , \quad \underline{u} \geq \underline{o}\} \tag{2.1.38}$$

(see, Luenberger (1973)) where the dual variable $\underline{u}$ is m_o-dimensional. The optimal solution of (2.1.38) is a feasible point of R (Luenberger (1973)). The applicaton of the simplex method to solve (2.1.38) illustrates the relationship of Algorithm (2.1.34) to the simplex method. A basic solution of (2.1.38) is obtained by partitioning N_{m_o} columnwise such that

$$N_{m_o} = \left[N_B \mid N_N \right] \tag{2.1.39}$$

where N_B is $n \times n$ and N_N is $n \times (m_o-n)$ and, as stated earlier, in this section $m_o > n$. Furthermore, the columns of N_B are chosen to be linearly independent. The elements of $\underline{b}_{m_o}$ and $\underline{u}$ are similarly partitioned. Hence,

$$\underline{b}_{m_o}^T = \left[\underline{b}_B^T , \underline{b}_N^T \right] \tag{2.1.40}$$

$$\underline{u}^T = \left[\underline{u}_B^T , \underline{u}_N^T \right] .$$

Clearly, N_B in (2.1.39) and $\underline{b}_B$ in (2.1.40) correspond to N and $\underline{b}$ respectively in (2.1.35). The constraints of (2.1.38) become

$$N_{m_o}\underline{u} = N_B\,\underline{u}_B + N_N\,\underline{u}_N = \underline{o}$$

and since the columns of N_B are linearly independent

$$\underline{u}_B = -\,N_B^{-1}\,N_N\,\underline{u}_N\,.$$

Substituting this in the objective function of (2.1.38), we obtain

$$\langle \underline{b}_{m_o}\,,\,\underline{u}\rangle = \langle \underline{b}_B\,,\,\underline{u}_B\rangle + \langle \underline{b}_N\,,\,\underline{u}_N\rangle$$

$$= -\langle \underline{b}_B\,,\,N_B^{-1}N_N\underline{u}_N\rangle + \langle \underline{b}_N\,,\,\underline{u}_N\rangle\,. \qquad (2.1.41)$$

Using (2.1.35)

$$\underline{x} = N_B^{-T}\,\underline{b}_B$$

thus, (2.1.41) becomes

$$\langle \underline{b}_{m_o}\,,\,\underline{u}\rangle = -\langle N_N^T\,\underline{x} - \underline{b}_N\,,\,\underline{u}_N\rangle\,. \qquad (2.1.42)$$

In the $m_o(m_o>n)$ dimensional optimisation problem (2.1.38), N_B is defined to be a basis since N_B consists of n linearly independent columns that can be regarded as a basis for the space E^n, which is the subspace of the variables eliminated from the objective function of (2.1.38). For the maximisation problem (2.1.38), the simplex method chooses the element of the dual vector $\underline{u}_N$ in (2.1.42) with a strictly positive coefficient. A positive coefficient in (2.1.42) signifies a violated constraint. Thus, the choice, in Rule (2.1.33) of a violated constraint normal to replace a current column of N corresponds choosing a positive coefficient of $\underline{u}_N$ in the simplex algorithm. The simplex method usually chooses the largest positive coefficient, however Rule (2.1.33) suggests an alternative criterion, as will be shown below. Let the chosen coefficient be that of the jth element of $\underline{u}_N$. Given this choice, the simplex method searches for the candidates to leave the basis (i.e. all the columns of N_B) in the jth column of

$$N_B^{-1} N_N . \qquad (2.1.43)$$

This column multiplies the jth element of $\underline{u}_N$. The columns of N_B corresponding to positive elements in the jth column of (2.1.43) are possible candidates. If $\underline{n}_j$ is the jth column of N_N, corresponding to the normal of the violated constraint that is the jth coefficient in (2.1.42), the jth column of (2.1.43) corresponds to

$$\underline{\lambda} = N_B^{-1} \underline{n}_j \qquad (2.1.44)$$

which is the vector computed in (2.1.36) for $N=N_B$, $\underline{n}_{j*} = \underline{n}_j$. If none of the elements of $\underline{\lambda}$ are strictly positive, the ith element of $\underline{u}_N$ in (2.1.42) can be increased indefinitely while improving the objective function value. This implies that if $\underline{\lambda} \leq o$ the dual problem (2.1.38) is unbounded; from which, by the Duality Theorem of Linear Programming (Luenberger (1973)), the infeasibility of the primal problem is concluded. This result is also proved in Theorem (2.4). Thus, the set Λ computed in Rule (2.1.31) is the set of elements of the jth column of (2.1.43) that correspond to the columns of N_B which can be exchanged by $\underline{n}_j$.

Hence, the selection of a violated constraint normal to replace an existing column of N_B and the selection of the column to be replaced on the basis of Rule (2.1.31) corresponds to the simplex algorithm applied to the dual problem (2.1.38). Among the violated constraint normals, the simplex algorithm normally chooses that of the most violated constraint (i.e. the largest positive coefficient of $\underline{u}_N$ in (2.1.42))to replace a column of N_B. Among the columns of N_B, to be replaced, the simplex algorithm normally chooses the ith column, where i solves

$$\min \; \{ \frac{c_i}{\lambda_i} \; | \; \lambda_i > o \} \qquad (2.1.45)$$

with c_i denoting the ith right hand of the equality constraints in (2.1.38). Clearly, $\lambda_i \in \Lambda$ with $\underline{n}_i$ replacing $\underline{n}$ in (2.1.32). In (2.1.38) all the right hand sides are zero. Bland (1977) has introduced a refinement to the criterion (2.1.45) in that the constraint normal chosen to leave the column of N_B is that satisfying (2.1.45) and that has the lowest index $k^*(i) \in J$ discussed in Rule (2.1.33). Since in (2.1.38) all the right hand elements of the equality constraints are zero, Bland's rule reduces to the lowest index criterion employed in Rule (2.1.33). Similarly, for the dual variable to enter the basis Bland (1977) suggests the choice of the element of $\underline{u}_N$, with a positive coefficient in (2.1.42), corresponding to the constraint with the lowest index $k \in J$. This is the same criterion employed in Rule (2.1.33) for choosing the column of N_B (i.e. N) to be replaced by the incoming violated constraint normal. Thus, Rule (2.1.33) is basically the same as Bland's rule except that in Bland's rule the coefficient c_i in (2.1.45) is not necessarily restricted to zero. Algorithm (2.1.34) is thus a method equivalent to a simplex algorithm for solving (2.1.38) with the refinements suggested by Bland (1977) for linear programming problems. Based on Bland's result, the convergence, in finite number of iterations, of Algorithm (2.1.34) may be stated. An iteration is taken here to be the steps during which the columns of N are unchanged, and ending in a single column of N being replaced by a violated constraint normal.

Theorem (2.3)

Algorithm (2.1.34) under Rules (2.1.31) and (2.1.33) converges in a finite number of iterations.

Proof

Due to the correspondence of Algorithm (2.1.34) under (2.1.31) and (2.1.33), and the simplex method under Bland's rule, which is basically Rule (2.1.33) as discussed above, this theorem is a restatement of Bland's result that the simplex method under his rule cannot cycle, hence it is finite (see Bland (1977), Theorem 1.1)). □

When the matrix N_{m_o} does not possess n linearly independent columns, the $n \times n$ nonsingular matrix N_B in (2.1.39) cannot be defined without further modification. Let N, be an $n \times \bar{n}$ ($\bar{n} < n$) matrix of linearly independent normals in N_{m_o}. Let this matrix be augmented by $n - \bar{n}$ columns of the $n \times n$ identity matrix I such that the columns of

$$N_B = \left[N_1 \;\vdots\; I_{n-\bar{n}} \right] \tag{2.1.46}$$

are linearly independent where $I_{n-\bar{n}}$ is the matrix of the selected columns of I. The selection of columns of I to be included in $I_{n-\bar{n}}$ may be done to ensure linear independence as each column is augmented to N_1 (see Section (3.5.1)). Let $\underline{e}$ be the $n - \bar{n}$ dimensional vector whose elements are all equal to 1 and let M be a large positive number, then (2.1.37) may be written as

$$\min \{ <\underline{o},\underline{x}> \mid N_{m_o}^T \underline{x} \geq \underline{b}_{m_o},\ I_{n-\bar{n}}^T \underline{x} \geq -M\,\underline{e} \}. \tag{2.1.47}$$

The dual of (2.1.47) is

$$\max \{ < \underline{u}, \underline{b}_{m_o} > - M < \underline{e}, \underline{\xi} > \mid N_{m_o} \underline{u} + I_{n-\bar{n}} \underline{\xi} = o; \underline{u}, \underline{\xi} \geq \underline{o} \} \tag{2.1.48}$$

where $\underline{u}, \underline{\xi}$ are the dual vectors associated with (2.1.47).

Thus Algorithm (2.1.34) may be used to compute a feasible point of the region described by the inequalities

$$N_{m_o}^T \underline{x} \geq \underline{b}_{m_o} \quad , \quad I_{n-\bar{n}}^T \underline{x} \geq - M \underline{e}$$

provided M is large enough, since N_B in (2.1.46) is nonsingular. When the number of linearly independent columns of N_{m_o} is less than n, a feasible point may also be found without introducing the extra bounds discussed above. A projection algorithm that does this by computing feasible points that need not be vertices is discussed in Section (2.2.2).

2.2 PROJECTION ALGORITHMS, REDUNDANCY AND DEGENERACY

The degeneracy and redundancy of constraints are discussed in Section (2.2.1). In Section (2.2.2) a projection algorithm is described for computing a feasible point of a linearly constrained region.

2.2.1 Degeneracy, Redundant Constraints, Infeasibility and Computational Considerations

In this Section $I(\underline{x})$ denotes the active constraints at $\underline{x}$, used in computing the point $\underline{x}$. The problem of degeneracy will be considered when at the point $\underline{x}$, computed using the $m \leq n$

constraints in $I(\underline{x})$, another constraint, not represented in $I(\underline{x})$, is also found to be satisfied as an equality constraint. Degeneracy occurs if the normal of this constraint say, $\hat{\underline{n}}$, is also linearly dependent on the normals of the constraints in $I(x)$, i.e.

$$N \hat{\underline{\lambda}} = \hat{\underline{n}} \qquad (2.2.1)$$

where the columns of N are the normals of the constraints in $I(\underline{x})$. The interpretation of this is that, in the linearly dependent case the new constraint is redundant in defining the intersection formed by the constraints in $I(\underline{x})$. However it need not be redundant for bounding the region R. For total redundancy of constraints Theorem (2.2) is required to hold. It was shown in Section (2.1.1) that once a satisfied constraint normal is linearly dependent on the normals of a current intersection, it will remain satisfied in this intersection.

The conventional method for determining the non-redundant constraints at an intersection is perturbing the right-hand side terms of the constraints active in this intersection (see, Gill and Murray (1974), Dantzig (1963)).

Consider also the case when a constraint with normal $\underline{n}$ is violated and has to be exchanged with one in $I(\underline{x}_f)$ but the largest element of $\underline{\lambda}$ computed using (2.1.11) or (2.1.12) is non positive. Thus

$$\underline{n} = N \underline{\lambda} \qquad (2.2.2)$$

where the columns of N are the normal vectors of the constraints in $I(\underline{x}_f)$, and

$$\underline{\lambda} \leq \underline{o} \,. \qquad (2.2.3)$$

The following result states that, in this case the region R is empty, implying infeasibility.

Theorem (2.4)

If

$$< \underline{n} , \underline{x}_f > - b < o \qquad (2.2.4)$$

and

$$N^T \underline{x}_f - \underline{b} = \underline{o} \qquad (2.2.5)$$

with

$$\underline{n} = N \underline{\lambda} , \quad \underline{\lambda} \le o \qquad (2.2.6)$$

then

$$\{ \underline{x} \in E^n \mid N^T \underline{x} - \underline{b} \ge \underline{o} \} \cap \{ \underline{x} \in E^n \mid < \underline{n} , \underline{x} > - b \ge o \} = \phi . \qquad (2.2.7)$$

Proof

Using (2.2.4) , (2.2.5) and (2.2.6) we have

$$b > < \underline{n} , \underline{x}_f > = < N \underline{\lambda} , \underline{x}_f > = < \underline{\lambda} , \underline{b} >$$

and for every $\underline{x}$ satisying $N^T \underline{x} \ge \underline{b}$,

$$< \underline{n} , \underline{x} > = < N\underline{\lambda} , \underline{x} > \le < \underline{\lambda} , \underline{b} > < b .$$

Hence

$$\{ \underline{x} \in E^n \mid < \underline{n} , \underline{x} > - b < o \} \supset \{ \underline{x} \in E^n \mid N^T \underline{x} - \underline{b} \ge \underline{o} \} \qquad (2.2.8)$$

and since the left hand term in (2.2.8) is the complement of the half space

$$h = \{ \underline{x} \in E^n \mid \langle \underline{n}, \underline{x} \rangle - b \geq o \}, \qquad (2.2.9)$$

(2.2.8) implies (2.2.7) . □

Theorem (2.2) established the condition under which a constraint may be redundant. The following result,which is in a sense the converse of Theorem (2.4),establishes another way in which redundancy may be detected.

Theorem (2.5)

If for R' defined as

$$R' \triangleq \{ \underline{x} \in E^n \mid N^T\underline{x} - \underline{b} \geq o \} , \qquad (2.2.10)$$

$$R' \cap \{ \underline{x} \in E^n \mid \langle \underline{n}, \underline{x} \rangle - b < o \} = \phi \qquad (2.2.11)$$

where $\underline{n}$ is not a column of N in R',

then the half space h given by (2.2.9),

$$h \supset R' \qquad (2.2.12)$$

and thus h is redundant.

Proof

The relationship (2.2.11) implies

$$h \cap R' = R' . \qquad (2.2.13)$$

Since the complement of h is

$$\{ \underline{x} \in E^n \mid < \underline{n} , \underline{x} > - b < o \} ,$$

using (2.2.11)

$$h \cap R' - R' = \{\underline{x} \in E^n \mid <\underline{n},\underline{x}> - b < o \} \cap R' = \phi \qquad (2.2.14)$$

is obtained. The result follows from (2.2.14) and (2.2.13). □

Given $\underline{x}_f$ as the current point in the intersection

$$N^T \underline{x} - \underline{b} = \underline{o} \qquad (2.2.15)$$

and the constraint described by the halfspace h (2.2. 9) and violated at $\underline{x}_f$,

$$<\underline{n},\underline{x}_f> - b < o , \qquad (2.2.16)$$

the formulae for updating the projection operators involved in (2.1.5) and (2.1.6) are discussed in Sections (3.5.1) - (3.5.2). These formulae compute updates when $\underline{n}$ is added as an extra column to N or an existing column of N is dropped. When rank (N)= n , the updating and the computation of successive projections take a simple form that is summarised below.

As mentioned in (2.1.8) when rank (N)= n ,

$$\underline{x}_f = (N^T)^{-1}\underline{b} , \qquad \underline{\lambda} = N^{-1}\underline{n} \qquad (2.2.17)$$

and the normal $\underline{n}$ in (2.2.16) can only be exchanged with a column of N. Let N have a column, say $\underline{n}_i$, for which $\lambda_i > o$, with $\underline{n} = N\underline{\lambda}$. Thus, $\underline{n}$ replaces $\underline{n}_i$ in N and, correspondingly, b in (2.2.16) replaces b_i in (2.2.17). Hence (2.2.17) may be recomputed using the new values for N and b, say $\hat{N}$ and $\hat{\underline{b}}$ respectively. The computation of $(\hat{N}^T)^{-1}$ may be done using $(N^T)^{-1}$, its ith column $\underline{\tilde{n}}_i$ and $\underline{n}$ with the formula

$$(N^T)^{-1} = (N^T)^{-1} - \frac{\underline{\tilde{n}}_i \, \underline{n}^T \, (N^T)^{-1}}{<\underline{\tilde{n}}_i \, , \underline{n}>} + \frac{\underline{\tilde{n}}_i \, \underline{e}_i^T}{<\underline{\tilde{n}}_i \, , \underline{n}>} \qquad (2.2.18)$$

where $\underline{e}_i$ is the column vector with 1 in its ith term and o elsewhere. Using its definition, the vector $\hat{\underline{b}}$ may be expressed as

$$\hat{\underline{b}} = \underline{b} + \underline{e}_i b - \underline{e}_i \, b_i \, . \qquad (2.2.19)$$

The new projection is computed with

$$\underline{x}_f' = (N^T)^{-1} \, \hat{\underline{b}}$$

$$= \left((N^T)^{-1} - \frac{\underline{\tilde{n}}_i \, \underline{n}^T (N^T)^{-1}}{<\underline{\tilde{n}}_i \, , \underline{n}>} + \frac{\underline{\tilde{n}}_i \, \underline{e}_i^T}{<\underline{\tilde{n}}_i , \underline{n}>} \right) (\underline{b} + \underline{e}_i b - \underline{e}_i b_i) \, . \qquad (2.2.20)$$

Since $\underline{e}_i^T \, \underline{b} = b_i$, $(N^T)^{-1} \underline{e}_i b = \underline{\tilde{n}}_i b$, $(N^T)^{-1} \underline{e}_i b_i = \underline{\tilde{n}}_i b_i$ (2.2.20) becomes

$$x_f' = (N^T)^{-1} \, \underline{b} - \frac{\underline{\tilde{n}} \, \underline{n}^T \, (N^T)^{-1} \, \underline{b}}{<\underline{\tilde{n}}_i \, , \underline{n}>} + \frac{\underline{\tilde{n}}_i \, b_i}{<\underline{\tilde{n}}_i , \underline{n}>}$$

$$+ \underline{\tilde{n}}_i \, b_i - \frac{\underline{\tilde{n}}_i \, \underline{n}^T \, \underline{\tilde{n}}_i \, b}{<\underline{\tilde{n}}_i \, , \underline{n}>} + \frac{\underline{\tilde{n}}_i \, b}{<\underline{\tilde{n}}_i , \underline{n}>}$$

$$- \underline{\tilde{n}}_i \, b_i + \frac{\underline{\tilde{n}}_i \, \underline{n}^T \, \underline{n}_i \, b_i}{<\underline{\tilde{n}}_i \, , \underline{n}>} - \frac{\underline{\tilde{n}}_i \, b_i}{<\underline{\tilde{n}}_i , \underline{n}>}$$

which, using (2.2.17), and $<\underline{\tilde{n}}_i , \underline{n}> = \lambda_i > o$ reduces to

$$\underline{x}_f' = \underline{x}_f + \alpha \underline{\tilde{n}}_i \qquad (2.2.21)$$

where

$$\alpha = \frac{-(<\underline{n}, \underline{x}_f> - b)}{\lambda_i} > 0 \qquad (2.2.22)$$

Thus the computation of successive projections to $\underline{x}_u$ using (2.2.17) is simplified to (2.2.21).

2.2.2 Projection Algorithms for Computing Feasible Points of a Linearly Constrained Region

In this section a few well known methods for obtaining a feasible point of a linearly constrained region are briefly discussed and a projection algorithm is suggested for solving the linear programming subproblem posed by these methods.

Perhaps the most popular method for calculating a feasible point involves the construction of the artificial objective function

$$\Psi(\underline{x}) = -\sum_{v \in V(\underline{x}_f)} (<\underline{n}_v , \underline{x}> - b_v) \qquad (2.2.23)$$

at an infeasible point $\underline{x}_f$, and the solution to

$$\min \{ \Psi(\underline{x}) \mid <\underline{n}_j , \underline{x}> - b_j \geq 0 , j \in S(\underline{x}_f) \cup I(\underline{x}_f)\} . \qquad (2.2.24)$$

This problem is equivalent to the phase one of the simplex algorithm in linear programming, started at $\underline{x}_f$ (see Dantzig (1963) pp. 101-119). The objective function (2.2.23) is sometimes called the infeasibility form and unless $\Psi(\underline{x}) = 0$ at the solution of (2.2.24) it will be concluded that the region R has no feasible solution. When $m_0 > n$, $\underline{x}_f$ (or $\underline{x}_u$) is usually chosen to be an infeasible vertex and the simplex method is applied which will ultimately compute a feasible vertex. However, in nonlinear programming the case $m_0 < n$ is frequently encountered. A vertex will not exist when $m_0 < n$ whereas feasible points may exist and can be located using projection methods. At $\underline{x}_f$, $V(\underline{x}_f) \neq \phi$, the steepest descent direction of (2.2.23) is given by

$$- \nabla\Psi(\underline{x}) = \sum_{v \in V(\underline{x}_f)} \underline{n}_v \ . \qquad (2.2.25)$$

This direction is projected into the intersection of the constraints in $I(\underline{x}_f)$, i.e.

$$- P \ \nabla\Psi(\underline{x}) \qquad (2.2.26)$$

is computed. The step taken along this projected direction may be restricted by the boundary of the feasible region thereby avoiding the transgression of a constraint in

$$K \triangleq \{ s \in S(\underline{x}_f) \mid - \langle \underline{n}_s, P \ \nabla\Psi(\underline{x}) \rangle < o \}. \qquad (2.2.27)$$

As an alternative strategy the step may be extended along this direction until the objective function (now incorporating the newly transgressed constraint(s) ceases to be reduced. Gill and Murray (1974) discuss a "non-simplex strategy" allowing the computation of a feasible point which is not necessarily a vertex. Fletcher (1970) also discusses the minimsation of (2.2.23). In his method an angle criterion is used to determine the optimum bound to add to the active set to form the initial vertex. This criterion is based on choosing the bound whose normal forms the largest angle with the subspace spanned by the normals $\underline{n}_i$, $i \in I(\underline{x}_u)$. Fletcher's algorithm starts with a vertex and follows a simplex strategy to generate subsequent vertices leading to a feasible one.

There are two other well known algorithms based on the simplex method. The first combines the two phases of the simplex method by augmenting the original objective function with (2.2.17) multiplied by a large positive number. Both functions are minimised at the same time. A non-simplex strategy may also be applied to this when the original objective function is nonlinear. The second algorithm solves the problem

$$\min \{ \xi \mid \langle \underline{n}_v , \underline{x} \rangle - b_v + \xi \geq 0, \quad v \in V(\underline{x}_u) \; ;$$

$$\langle \underline{n}_i , \underline{x} \rangle - b_i \geq 0 , \; i \in I(\underline{x}_u) \cup S(\underline{x}_u) \}$$

for which $\underline{x}_u$ and ξ^o may be regarded as starting points with ξ^o chosen sufficiently large (see, e.g. Zoutendjik (1970)).

There are also a number of relaxation methods for computing a feasible point of linearly constrained region. The method due to Motzkin and Schoenberg (1954) (see also Herman (1975)) uses the current point $\underline{x}_f$ and one of the constraints it violates to generate the next point

$$\underline{x}_{f'} = \underline{x}_f + \alpha(\underline{x}_f^P - \underline{x}_f)$$

where $\underline{x}_f^P$ is the orthogonal projection of $\underline{x}_f$ onto the chosen violated constraint (i.e. (2.1.5) with $\underline{x}_f$ replacing $\underline{x}_u$ and N and $\underline{b}$ defined by the normal and right hand side of the violated constraint). The scalar α is chosen from $0 < \alpha \leq 2$ with $0 < \alpha < 1$ termed as "over-relaxation" $\alpha = 1$ as "projection" and $\alpha = 2$ as "reflexion". It should be noted that no criterion for choosing the violated constraint is utilised in the convergence proofs. The case $\alpha = 1$ is discussed in detail by Agmon (1954). Both these methods either generate infinite sequences or converge to a feasible point but do not have a means of identifying an infeasible region. Also, they are normally not expected to converge to vertices. Since methods based on projections and the simplex method, without cycling, guarantee convergence in finite number of steps, they are generally superior to relaxation methods.

Finally, a projection algorithm based on the minimisation of an infeasibility form is given below for a general projection operator.

Algorithm (2.2.28)

Step 0: Given $\underline{x}_k$, set $\underline{x}_f = \underline{x}_k$

Step 1: Set up $V(\underline{x}_f)$ (or delete from it any constraint satisfied during the previous iteration)

If $V(\underline{x}_f) = \phi$, $\underline{x}_f$ is feasible, stop. Otherwise set up $\Psi(\underline{x})$ in (2.2.23) and $\nabla\Psi(\underline{x})$ in (2.2.25).

Step 2: Select the constraints satisfied as equalities at $\underline{x}_f$, construct (or update) the matrix N whose columns are the normals of these constraints. Assuming degeneracy will not occur, N will have at most n columns. Compute $\underline{\lambda}$ satisfying

$$N\underline{\lambda} = - \nabla\Psi(\underline{x}) \qquad (2.2.29)$$

If an element of $\underline{\lambda}$ is strictly positive, delete the corresponding column of N. (Let this deleted column be the normal of constraint i).

Set up (or update by deleting constraint i) the set $I(\underline{x}_f)$ from the constraints whose normals are still in matrix N. If $\text{rank}(N) = n$ and $V(\underline{x}_f) \neq \phi$, there is no feasible solution : stop (see Theorem (2.4)).

Step 3: Set up (or update using constraint i in Step 2 or constraint k* in Step 4, from the previous iteration) the projection operator P and compute the projection (2.2.26).

Set up $S(\underline{x}_f)$ (including constraint i from Step 2)

Step 4: Compute

$$\alpha_{k^*} = \min_k \left\{ \alpha_k = \frac{(\langle \underline{n}_k, \underline{x}_f \rangle - b_k)}{\langle \underline{n}_k, P\, \nabla\Psi(\underline{x}) \rangle},\ k \in K \right\} \tag{2.2.30}$$

where K is given by (2.2.27) and k* is the index of the constraint corresponding to α_{k^*} .

Step 5: Reset $\underline{x}_f$ to

$$\underline{x}_f = \underline{x}_f - \alpha_{k^*}\, P\nabla\Psi(\underline{x}) \tag{2.2.31}$$

and go to Step 1.

In Step 4, when there is more than one constraint corresponding to k* , they may be added to the active set as long as the columns of N in Step 2 remain linearly independent. In the linearly dependent case the redundancy and degeneracy discussion in Section (2.2.1) applies. When the feasible region is unbounded in the direction $P\,\nabla\,\Psi(\underline{x})$, α_{k^*} may be set to the value for which $\underline{x}_f$ in (2.2.31) satisfies all the violated constraints.

There are a number of alternatives for the projection operator P. It can be

$$P = Z(Z^TZ)^{-1}\, Z^T \tag{2.2.32,a}$$

where the matrix Z is defined in Section (1.1.4) such that $Z^TN = 0$,

or,

$$P = Z(Z^TGZ)^{-1}\, Z^T \tag{2.2.32,b}$$

where G is a symmetric positive definite matrix, or

$$P = I - N(N^TN)^{-1}\, N^T \tag{2.2.32,c}$$

or,

$$P = I - G^{-1}N(N^TG^{-1}N)^{-1}\, N^T\ . \tag{2.2.32,d}$$

In (2.2.31) $\nabla\Psi$ is replaced by $G\,\nabla\Psi$ for (2.2.32,b) and by $G^{-1}\nabla\Psi$ for (2.2.32,d). Note that $-G\,\nabla\Psi$ and $-G^{-1}\nabla\Psi$ are still descent directions. An alternative choice for P is suggested by Gill and Murray (1974) who take $P = ZZ^T$. The choice for P from the alternatives in (2.2.32) will depend on the nature of the quadratic or nonlinear programming algorithm in connection to which a method to calculate feasible points is required.

The convergence of Algorithm (2.2.28) follows from the reduction of the objective function every time the step (2.2.31) is taken. The degenerate case, as mentioned earlier, may be resolved by perturbing the constraints. In the non-degenerate case, zero step lengths ($\alpha_{k^*} = o$) are not possible since a constraint dropped from the active set, and hence from the columns of N in Step 2, has a positive multiplier λ_i.

A constraint with normal $\underline{n}_i$ is deleted from the column of N in (2.2.29) to yield $\hat{N}$, when $\lambda_i > o$. The Projection operator P is accordingly updated so that $P\hat{N} = o$ (for (2.2.32,d) $PG^{-1}\hat{N} = o$), thus

$$\hat{N}\underline{\lambda} + \lambda_i\,\underline{n}_i = -\nabla\Psi(\underline{x}) \qquad (2.2.33)$$

where $\underline{\lambda}$ is the vector in (2.2.29) without its ith element. Premultiplying this equation by P (or PG^{-1} for (2.2.32,d)) yields

$$\lambda_i\,P\,\underline{n}_i = -P\,\nabla\Psi(\underline{x}) \qquad (2.2.34)$$

thus

$$o \le \lambda_i\,\langle \underline{n}_i\,,\,P\,\underline{n}_i\rangle = -\langle \underline{n}_i\,,\,P\,\nabla\Psi(\underline{x})\rangle \qquad (2.2.35)$$

or for (2.2.32,d)

$$o \le \lambda_i\,\langle \underline{n}_i\,,\,PG^{-1}\,\underline{n}_i\rangle = -\langle \underline{n}_i\,,\,PG^{-1}\,\nabla\Psi(\underline{x})\rangle \qquad (2.2.36)$$

since P (or PG^{-1}) is a symmetric projection operator. Thus, the direction $- P \nabla\Psi(\underline{x})$ (or $-PG^{-1} \nabla\Psi(\underline{x})$) satisfies the constraint i. Until a violated constraint is satisfied and thus dropped from the objective function and its gradient, the direction $- P \nabla\Psi(\underline{x})$ (or $PG^{-1} \nabla\Psi(\underline{x})$) will therefore satisfy i . If a violated constraint is deleted and a new gradient $\nabla\Psi(\underline{x})$ computed, the descent direction $- P \nabla\Psi(\underline{x})$ may again lead to i. Thus an intersection can only recur when a violated constraint is satisfied in the meantime. As there are a finite number of intersections and a finite number of violated constraints, convergence occurs in a finite number of steps (i.e., a finite number of changes in the active set and the satisfaction of a finite number of violated constraints).

The choice of the constraint to be deleted from $I(\underline{x}_f)$ may be done in two ways. The first is choosing the constraint corresponding to the largest positive element of $\underline{\lambda}$ in (2.2.29). Premultiplying (2.2.33) with $- P \nabla\Psi(\underline{x})$,using $P\hat{N} = o$, this choice may be interpreted as

$$\lambda_i \; < \underline{n}_i \; , P \nabla\Psi(\underline{x}) > \; = \; - \; < \nabla\Psi(\underline{x}), P \nabla\Psi(\underline{x}) >$$

and so deleting constraint i gives the smallest directional derivative per unit change in the residual $< \underline{n}_i \; , P \nabla\Psi(\underline{x}) >$. A similar result may be obtained using (2.2.32,d) with PG^{-1}. Another strategy has been suggested by Goldfarb and Reid (1975) in which the constraint giving the largest descent direction per unit step is chosen. Hence the constraint maximising

$$\frac{< \nabla\Psi(\underline{x}), \; P \nabla\Psi(\underline{x}) >}{< P\nabla\Psi(\underline{x}), \; P \nabla\Psi(\underline{x}) >^{\frac{1}{2}}} \qquad (2.2.37)$$

is chosen. Note that $P^T P = P$, and P is the operator obtained after deleting the constraint being tested from the active set and

$- P \nabla\Psi(\underline{x})$ is the resulting search direction. After setting $P^T P = P$, (2.2.37) yields

$$< \nabla\Psi(\underline{x}), P \nabla\Psi(\underline{x}) >^{\frac{1}{2}}$$

and using (2.2.34), this becomes

$$\lambda_i \quad < \underline{n}_i, P\underline{n}_i >^{\frac{1}{2}} . \qquad (2.2.38)$$

Thus the constraint i that maximises (2.2.38) is chosen to be dropped. If P is chosen to be (2.2.32,b) or (2.2.32,d) the relationship $P^T P = P$ used above takes the forms $P^T G P = P$ and $(PG^{-1})^T G (PG^{-1}) = PG^{-1}$ respectively. For (2.2.32,b) the denominator of (2.2.37) is redefined as a "weighted" product $< P\nabla\Psi(\underline{x}), G P \nabla\Psi(\underline{x}) >^{\frac{1}{2}}$ and the result once more turns out to be (2.2.38). For (2.2.32,d) the same denominatior takes the form $< P G^{-1} \nabla\Psi(\underline{x}), G P G^{-1} \nabla\Psi(\underline{x}) >^{\frac{1}{2}}$ and the condition is replaced by $\lambda_i < \underline{n}_i , PG^{-1} \underline{n}_i >^{\frac{1}{2}}$.

2.3 CONCLUDING REMARKS

Algorithms (2.1.34) and (2.2.28) illustrate the connection of feasible point algorithms to projection methods in quadratic and nonlinear programming. In particular, the unified use of projection operators is demonstrated. The convergence of both algorithms are discussed. The relationship of infeasible points and their projections to redundant constraints and infeasible regions are discussed in Sections (2.1.2) and (2.2.1).

Avis and Chvatal (1978) discuss the numerical performance of the pivoting rule, (2.1.33) in Section (2.1.2), due to Bland (1977). On randomly generated linear programming problems with 50 nonnegative variables and 50 additional inequalities Bland's rule is reported to have required about 400 iterations of the revised simplex method (see, Luenberger (1973)). The corresponding figure for the rule choosing the largest coefficient in the objective function (in maximisation problems) is about 100 iterations. However, the unique feature of Bland's rule is that it excludes the possibility of cycling.

CHAPTER 3

AN ALGORITHM FOR POSITIVE DEFINITE QUADRATIC PROGRAMMING

3.1 INTRODUCTION

In this chapter a method is developed for solving positive definite quadratic programming problems. Such problems are special cases of quadratic programming in which the second derivative matrix of the quadratic objective function is assumed to be positive definite. The present method uses the unconstrained minimum, $\underline{x}_u$, of the objective function to compute the global constrained optimum. This involves the projection of $\underline{x}_u$ onto the feasible region.

Positive definite quadratic programming problems occur in a number of areas of applied mathematics. One such field is optimal control where quadratic objective functions are widely used (see, Cannon, Cullum, Polak (1970), Polak (1971)). The algorithm discussed in this chapter poses general positive definite quadratic programming problems as equivalent quadratic programs with the transformed objective function

$$< \underline{x} - \underline{x}_u \, , \, G \, (\underline{x} - \underline{x}_u) > \; .$$

Objective functions of this structure are common in optimal control where $\underline{x}$ is the transcription of the trajectory of a dynamic system into static form and $\underline{x}_u$ is the fixed desired path the optimal solution is required to follow. Such formulations will be discussed in Chapters 5 and 6. The algorithm in this chapter is in particular intended to solve the positive definite quadratic programming subproblems generated by the algorithm in Chapter 4.

Preliminary discussion on the projections used by the algorithm is presented in Section (3.2). Some properties of the quadratic objective function and inequality constraints are discussed in Section (3.3). In Section (3.4) the algorithm is introduced. The recurrance relationship for updating projection operators when the algorithm adds or drops constraints from the active set are discussed in Section (3.5). In Section (3.6) the convergence properties of the algorithm are discussed.

3.1.1 The Unconstrained Minimum

Consider the problem

$$\min \{ q(\underline{x}) \mid \underline{x} \in E^n \} \tag{3.1.1}$$

with

$$q(\underline{x}) = \langle \underline{a}, \underline{x} \rangle + \tfrac{1}{2} \langle \underline{x}, G\,\underline{x} \rangle . \tag{3.1.2}$$

The point $\underline{x}_u$ which solves (3.1.1) must satisfy the necessary condition

$$\nabla q\,(\underline{x}_u) = \underline{o} \tag{3.1.3}$$

and for $\underline{x}_u$ to be unique, the matrix G is required to be positive definite (see, Luenberger (1973) p. 112). This implies that G is nonsingular, $q(\underline{x})$ is convex and

$$\underline{x}_u = -\,G^{-1}\,\underline{a} = -\,H\,\underline{a} \tag{3.1.4}$$

where H is simply used to denote G^{-1}.

3.1.2 The Quadratic Programming Problem

Consider the constrained optimisation problem

$$\min \{ q(\underline{x}) \mid \underline{g}(\underline{x}) \geq \underline{o} \} \tag{3.1.5}$$

where $\underline{g}$ is given by (1.1.2,b). A point $\underline{x}_c$ that solves (3.1.5), satisfies the necessary Kuhn-Tucker conditions

$$(G\,\underline{x}_c + \underline{a}) - \nabla\underline{g}(\underline{x}_c)\,\underline{\lambda}_c = \underline{o} \tag{3.1.6,a}$$

$$\underline{g}(\underline{x}_c) \geq \underline{o} \tag{3.1.6,b}$$

$$\langle \underline{\lambda}_c , g(\underline{x}_c) \rangle = 0 \tag{3.1.6,c}$$

$$\underline{\lambda}_c \geq \underline{o} \tag{3.1.6,d}$$

where $\underline{\lambda}$ is the m_o-dimensional Kuhn-Tucker multiplier defined in (1.1.9) and $\nabla\underline{g}$ is the $(n \times m_o)$ matrix of constraint normals (1.1.16) at $\underline{x}_c$. The matrix **G** is taken to be positive definite. Thus, under the assumption of concavity of the elements of $g(\underline{x})$, the conditions (3.1.6) are also sufficient for the global optimality of $\underline{x}_c$ (see, e.g. Fiacco and McCormick (1968) Theorem 18, or Luenberger (1973) p.235). This convexity assumption is not unrealistic since the constraints in (3.1.5) are taken to be linear inequalities throughout most of this chapter with (3.1.5) taking the special form

$$\min \{ q(\underline{x}) \mid N_{m_o}^T \underline{x} \geq \underline{b}_{m_o} \} \tag{3.1.7}$$

The matrix N_{m_o} is $(n \times m_o)$ dimensional, $\underline{b}_{m_o}$ is an m_o-dimensional vector and (3.1.7) is the quadratic programming problem.

Clearly, if the unconstrained optimum of $q(\underline{x})$ also satisfies the inequality constraints in (3.1.5), i.e. $\underline{g}(\underline{x}_u) \geq \underline{o}$, then $\underline{x}_u$ is also the solution of (3.1.5). It will be shown in Theorem (3.2) that if $\underline{x}_u$ violates any of the constraints, then the solution of (3.1.5) lies on the boundary of the feasible region

$$R \triangleq \{ \underline{x} \in E^n \mid \underline{g}(\underline{x}) \geq \underline{o} \} . \tag{1.1.22}$$

When the constraints are linear inequalities as in (3.1.7), R will be defined in the modified form

$$R \triangleq \{ \underline{x} \in E^n \mid N_{m_o} \underline{x} \geq b_{m_o} \} . \tag{3.1.8}$$

3.2 PROJECTION OPERATORS

In this section some properties of projection operators introduced in Section (1.1.2) will be briefly discussed. The constraints are assumed to be linear inequalities and the projection operators are computed to project vectors in E^n onto the subspace spanned by m linearly independent normals of the active constraints (see, Assumption (1.2), Section (1.1.2)).

3.2.1 Preliminaries

Consider the elements of the subspace Ω_o (1.1.34) in Definition (1.1) in Section (1.1.2). The (n - m) columns of Z (1.1.45) were defined as the basis vectors of Ω_o where $m \leq \min(n, m_o)$. Any vector $\underline{x} \in E^n$ may be expressed in terms of the two orthogonal components

$$\underline{x} = Z \underline{v} + N_m \underline{\lambda} \tag{3.2.1}$$

since $N_m^T Z = o$. The vectors $\underline{v}$ and $\underline{\lambda}$ are respectively n - m and m dimensional. A slightly different alternative to (3.2.1) is obtained when the elements of $\bar{\Omega}$ are used instead of $N_m\underline{\lambda}$. With $H = G^{-1}$ this yields

$$\underline{x} = Z\,\underline{v} + H\,N_m\,\underline{\lambda} \tag{3.2.2}$$

where now

$$<(Z\underline{v})\,,\,G\,(H\,N_m\,\underline{\lambda})> \;=\; o \tag{3.2.3}$$

and the orthogonality of the two components in (3.2.2) is defined with respect to G. Hence,

$$<\underline{x}_1\,,\,G\,\underline{x}_2> \;=\; o\;, \quad \forall\,\underline{x}_1 \in \Omega_o\,,\;\forall\,\underline{x}_2 \in \bar{\Omega}\,. \tag{3.2.4}$$

3.2.2 Projections in E^n

Consider the projections of $\underline{x} \in E^n$ onto Ω_o and $\bar{\Omega}$. Using (1.1.37) $\underline{x}$ can be written as

$$\underline{x} = \underline{x}_1 + \underline{x}_2\;, \quad \underline{x}_1 \in \Omega_o\,,\;\underline{x}_2 \in \bar{\Omega}\,. \tag{3.2.5}$$

Since $\underline{x}_2 = H\,N_m\,\underline{\lambda}$, for some m-vector λ, and since $N_m^T\underline{x}_1 = \underline{o}$ by (1.1.34),

$$N_m^T\,\underline{x} = N_m^T\,\underline{x}_2 = (N_m^T\,H\,N_m)\,\underline{\lambda}\,,$$

hence

$$\underline{x}_2 = H\,N_m\,(N_m^T\,H\,N_m)^{-1}\,N_m^T\,\underline{x} \tag{3.2.6}$$

and using (3.2.5),

$$\underline{x}_1 = (I - H\,N_m\,(N_m^T\,H\,N_m)^{-1}\,N_m^T)\,\underline{x} \tag{3.2.7}$$

is obtained. Setting $P = (I - H\,N_m\,(N_m^T\,H\,N_m)^{-1}\,N_m^T)$ the relationships

$P\,\underline{x}_2 \;=\; P\,H\,N_m\,\underline{\lambda} \;=\underline{o}$, $P\,\underline{x}_1 = \underline{x}_1$, $P\,P \;=\; P$, $(I - P)\,P \;=\; o$, $(I - P)\,\underline{x}_1 = o$, $(I - P)\,\underline{x}_2 = \underline{x}_2$ and $(I - P)(I - P) = (I - P)$ may be easily verified. Clearly, P projects $\underline{x} \in E^n$ onto Ω_o along $\bar{\Omega}$ and those vectors of the form $H\,N\,\underline{\lambda}$ into $\underline{o}$. Similarly, $(I - P)$ projects $\underline{x} \in E^n$ onto $\bar{\Omega}$ along Ω_o into a vector of the form $H\,N\,\underline{\lambda}$ and all and only those vectors in Ω_o into $\underline{o}$.

The projection of $\underline{x}$ onto the manifold Ω (1.1.35) may be obtained using (3.2.5) and $N_m^T\,\underline{x}_1 \;=\; \underline{b}_m$. Thus,

$$N_m^T\,\underline{x} \;=\; N_m^T\,\underline{x}_1 \;+\; (N_m^T\,H\,N_m)\underline{\lambda}$$

and

$$\underline{x}_2 \;=\; H\,N_m\,(N_m^T\,H\,N_m)^{-1}\,(N_m^T\,\underline{x} - \underline{b}_m) \qquad (3.2.8)$$

$$\underline{x}_1 \;=\; \underline{x} - H\,N_m(N_m^T\,H\,N_m)^{-1}\,(N_m^T\,\underline{x} - \underline{b}_m)\;. \qquad (3.2.9)$$

□

An alternative approach to the above projections is the representation $Z\,\underline{v} = \underline{x}_1$ with $Z^T G\,\underline{x}_2 = \underline{o}$ using (3.2.3). Premultiplying by $Z^T G$, (3.2.5) becomes

$$Z^T G\,\underline{x} \;=\; Z^T G\,\underline{x}_1 \;=\; Z^T G\,Z\,\underline{v}$$

for some $(n - m)$-vector $\underline{v}$. Hence

$$\underline{v} = (Z^T G\,Z)^{-1}\,Z^T\,G\,\underline{x}$$

and

$$\underline{x}_1 = Z(Z^T G\,Z)^{-1}\,Z^T G\,\underline{x}\,, \qquad (3.2.10)$$

$$\underline{x}_2 = (I - Z(Z^T G\,Z)^{-1}\,Z^T G)\,\underline{x}\;. \qquad (3.2.11)$$

The correspondance between $(I - P)$ and $(I - Z(Z^T G\,Z)^{-1}\,Z^T G)$, and, P and $Z(Z^T G\,Z)^{-1}\,Z^T G$ can be seen by comparing (3.2.7) with (3.2.10) and (3.2.6) with (3.2.11).

3.3 MOTIVATION FOR A POSITIVE DEFINITE QUADRATIC PROGRAMMING ALGORITHM

In this section solutions to the quadratic programming problem are outlined. These are based on the projection of the unconstrained minimum of the objective function onto the feasible region. The basic relationship between the unconstrained optimum of a quadratic function and its constrained optimum is given by Theorem (3.1). Further discussions related to the constraints violated at the unconstrained optimum are given in Section (3.3.2). A numerical example is given in Section (3.3.2).

3.3.1 The Quadratic Objective Function

In the presence of only $m \le n$ linear equality constraints, the quadratic optimisation problem (3.1.5) reduces to

$$\min \{ q(\underline{x}) \mid N_m^T \underline{x} = \underline{b}_m \} \qquad (3.3.1)$$

The optimality conditions (3.3.6) may be simplified for (3.3.1) and reduce to

$$\nabla q(\underline{x}_c) = N_m \underline{\lambda}_c \qquad (3.3.2)$$

$$N_m^T \underline{x}_c = \underline{b}_m$$

where the columns of N_m are assumed to be linearly independent.

For a quadratic function, the relationship

$$\underline{x}_c - \underline{x}_u = H(\nabla q(\underline{x}_c) - \nabla q(\underline{x}_u))$$

holds exactly. Using (3.1.3) and (3.3.2), this reduces to

$$\underline{x}_c = \underline{x}_u + H N_m \underline{\lambda}_c . \qquad (3.3.4)$$

Premultiplying (3.3.4) by N_m^T and applying (3.3.3) yields

$$N_m^T \underline{x}_c = \underline{b}_m = N_m^T \underline{x}_u + N_m^T H N_m \underline{\lambda}_c .$$

Thus,

$$\underline{\lambda}_c = -(N_m^T H N_m)^{-1} (N_m^T \underline{x}_u - \underline{b}_m), \qquad (3.3.5)$$

$$\underline{x}_c = \underline{x}_u - H N_m (N_m^T H N_m)^{-1} (N_m^T \underline{x}_u - \underline{b}_m) . \qquad (3.3.6)$$

Comparing (3.3.5) with (3.3.5) with (3.2.9), it may be concluded that $\underline{x}_c$, the solution of (3.3.1), is the projection of $\underline{x}_u$ onto the linear manifold Ω given by (1.1.35). When $\underline{b}_m = \underline{o}$, (3.3.6) may be compared with (3.2.7). In this case, (3.3.6) projects the unconstrained optimum of $q(\underline{x})$ onto Ω_o.

The projection of $\underline{x}_u$ on the convex region R is equivalent to the solution of the minimum norm problem, (see, Luenberger (1969), Chapter 3),

$$\min \{ \| \underline{x} - \underline{x}_u \|_G^2 \mid \underline{x} \in R \} \qquad (3.3.7)$$

where R is the convex region given by (1.1.22) or, for only linear constraints, by (3.1.8). The projection in (3.3.7) is with respect to the norm $\|\underline{y}\|_G^2 = \langle \underline{y}, G\,\underline{y} \rangle$. The link between the constrained optimisation problems (3.3.7) and (3.1.5) or (3.1.7) is summarised in the following theorem.

Theorem (3.1)

If $\underline{x}_c$ is a solution of the minimum norm problem (3.3.7), then it is also a solution of the quadratic optimisation problem (3.1.5).

Proof

Clearly, the optimal solution of (3.3.7) is not altered if this problem is modified to

$$\min \{ \tfrac{1}{2}\| \underline{x} - \underline{x}_u \|_G^2 \mid \underline{x} \in R \}. \tag{3.3.8}$$

The key to the proof is that the objective function in (3.3.8) is, in fact, equivalent to $q(\underline{x})$, except a constant part, if $\underline{x}_u=-H\underline{a}$ is substituted in this function. Conversely, if $\underline{x}$ is replaced by the transformation $\underline{x} - \underline{x}_u$, $\underline{x}_u=-H\underline{a}$, in $q(\underline{x})$, the objective function of (3.3.8) is obtained with an additional constant term. Since constant terms do not affect the optimal solution $\underline{x}_c$, the result follows. Also, if the optimality condition (3.1.6,a) is written for both (3.3.8) and (3.1.5),

$$G\,(\underline{x}_c - \underline{x}_u) - \nabla \underline{g}(\underline{x}_c)\,\underline{\lambda}_c = \underline{o} \tag{3.3.9}$$

$$\nabla q(\underline{x}_c) - \nabla \underline{g}(\underline{x}_c)\,\underline{\lambda}_c = \underline{o} \tag{3.3.10}$$

are obtained respectively. Since the relationship

$$\nabla q(\underline{x}_c) = \nabla q(\underline{x}_u) + G\,(\underline{x}_c - \underline{x}_u) \tag{3.3.11}$$

holds exactly, (3.1.3) may be applied to (3.3.11) to show that (3.3.10) is identical to (3.3.9). Thus, since the same optimality conditions apply for both (3.3.8) and (3.1.5), $\underline{x}_c$ solves both these problems. □

It should be noted that the feasible region is not required to be convex for the above result. In the nonconvex case both (3.3.8) and (3.1.5) have the same local constrained optima.

The generalisation of the solution (3.3.6) of (3.3.1) to linear inequality constrained problems, i.e. (3.1.7), may be obtained by computing projections of $\underline{x}_u$ onto the intersections of succesive active sets using (3.3.6). The Kuhn-Tucker multipliers for each active set are given by (3.3.5).

For computational efficiency and accuracy, the matrix N_m can be factorised as suggested by Gill and Murray (1978)

$$N_m^T = \left[L \mid 0 \right] Q \tag{1.1.79}$$

where L is an (m x m) lower triangular matrix and Q an (n x n) orthogonal matrix (i.e. $Q^T Q = Q\,Q^T = I$). Furthermore, Q is partitioned with the (m x n) submatrix Q_1 denoting the first m rows and the ((n - m) x n) submatrix Q_2 denoting the last n - m rows, thus

$$Q = \begin{bmatrix} Q_1 \\ \hline Q_2 \end{bmatrix} . \tag{1.1.80}$$

Hence, (3.3.5) and (3.3.6) may be expressed in factorised form using (1.1.79) and (1.1.80)

$$L^T \underline{\lambda}_c = (Q_1\, H\, Q_1^T)^{-1}\, L^{-1}\, (N_m^T\, \underline{x}_u - \underline{b}_m) \tag{3.3.12}$$

or

$$\underline{\lambda}_c = L^{-T}(Q_1\, H\, Q_1^T)^{-1}\, L^{-1}(N_m^T\, \underline{x}_u - \underline{b}_m) \tag{3.3.13}$$

and

$$\underline{x}_c = \underline{x}_u - H\, Q_1^T\, L^T\, \underline{\lambda}_c \; . \tag{3.3.14}$$

Furthermore, the $(\bar{L}\, D\, \bar{L}^T)$ Cholesky factors of $(Q_1\, H\, Q_1^T)$ may be used for computing (3.3.13) and (3.3.14). When a constraint is added to or dropped from the active set, the corresponding modifications to Q_1, Q_2, L and the Cholesky factors are discussed by Gill et.al. (1974) and Gill and Murray (1978).

It should be noted that in (3.3.5), (3.3.6) and (3.3.12), (3.3.13), the vector

$$N_m^T \underline{x}_u - \underline{b}_m \tag{3.3.15}$$

is only computed once since

$$N_{m_o}^T \underline{x}_u - \underline{b}_{m_o} \tag{3.3.16}$$

is computed once. Moreover, in contrast to well known algorithms (e.g. Fletcher (1971), Goldfarb (1972), Gill and Murray (1978)) the expressions (3.3.5), (3.3.6) and (3.3.12), (3.3.14) avoid altogether the computation of the gradient of the objective function. A saving in computation is thus obtained by solving (3.3.8) rather than (3.1.7).

Some existing quadratic programming algorithms may also be modified to adopt (3.3.8) instead of the original quadratic programming problem (3.1.7). The algorithms of Gill and Murray (1978), Goldfarb (1972) and Fletcher (1971) compute a sequence of subproblems

$$\min \{ <\underline{d}, G\,\underline{x}_k + \underline{a}> + \tfrac{1}{2}<\underline{d}, G\,\underline{d}> \mid N_m^T \underline{d} = \underline{o} \} \tag{3.3.17}$$

where $\underline{x}_k$ is any feasible point such that

$$N_{m_o}^T \underline{x}_k \geq \underline{b}_{m_o} \quad \text{and} \quad N_m^T \underline{x}_k = \underline{b}_m$$

and $\underline{x}_k + \underline{d}$ solves the equality constrained problem (3.3.1). The subproblem (3.3.17) is obtained by substituting $\underline{x}_k + \underline{d}$ in (3.3.1). Since the unconstrained optimum of the quadratic objective function in (3.1.17) is

$$\underline{d}_u = -H(G\,\underline{x}_k + \underline{a}) = -(\underline{x}_k - \underline{x}_u), \tag{3.3.18}$$

using Theorem (3.1) the solution of (3.1.17) is the same as the solution of

$$\min \{ \| \underline{d} - \underline{d}_u \|_G^2 \mid N^T \underline{d} = \underline{o} \} . \qquad (3.3.19)$$

All vectors that satisfy $N^T\underline{d} = \underline{0}$ may be expressed as $\underline{d} = Z\,\underline{v}$, using the arguments in Sections (3.2.1) and (3.2.2), with Z given by (1.1.45), $N^T Z = o$, and $\underline{v}$ an (n-m)-vector. Thus, (3.3.19) may be reduced to the (n-m) dimensional unconstrained optimisation problem

$$\min \{ \| Z\underline{v} - \underline{d}_u \|_G^2 \mid \underline{v} \in E^{n-m} \} . \qquad (3.3.20)$$

Since G is positive definite, the solution of (3.3.20) is similar to (1.1.49), hence

$$\nabla_v \| Z\underline{v} - \underline{d}_u \|_G^2 = 2\, Z^T G(Z\underline{v} - \underline{d}_u) = \underline{o} , \qquad (3.3.21)$$

$$(Z^T G\, Z)\, \underline{v} = Z^T G\, \underline{d}_u ,$$

thus,

$$\underline{d} = Z(Z^T G\, Z)^{-1} Z^T G\, \underline{d}_u \qquad (3.3.22)$$

or, substituting (3.3.18)

$$\underline{d} = - Z(Z^T G\, Z)^{-1} Z^T G(\underline{x}_k - \underline{x}_u) . \qquad (3.3.23)$$

Using (3.3.2), the Lagrange multipliers at $\underline{x}_k, \underline{\lambda}_k$, may be expressed as the solution of

$$N_m \underline{\lambda}_k = \nabla q(\underline{x}_k) ,$$

hence

$$\underline{\lambda}_k = (N_m^T N_m)^{-1} N_m^T\, G(\underline{x}_k - \underline{x}_u) \qquad (3.3.24)$$

or, using Businger and Golub's (1965) suggestion for computing the pseudo inverse $(N_m^T N_m)^{-1} N_m^T$ in terms of Q_1 and L,

$$\underline{\lambda}_k = L^{-T} Q_1\, G(\underline{x}_k - \underline{x}_u) .$$

The vector $\underline{d}$ in (3.3.23) is identical to that computed by the Gill and Murray (1978) algorithm when (3.3.19) is solved instead of (3.3.17). Gill and Murray (1978) compute $\underline{d}$ using

$$\underline{d} = - Z(Z^T G Z)^{-1} Z^T(G \underline{x}_k + \underline{a}) . \qquad (3.3.25)$$

Expression (3.3.23) is a more economical way of computing $\underline{d}$ rather than (3.3.25) when $\underline{x}_u$ and the product $Z^T G$ are readily available.

Another way of solving (3.3.19) is solving the optimality conditions of (3.3.19), i.e.

$$G(\underline{d} - \underline{d}_u) = N_m \underline{\lambda}_k \qquad (3.3.26)$$

$$N_m^T \underline{d} = \underline{o} . \qquad (3.3.27)$$

These equations yield

$$\underline{\lambda}_k = - (N_m^T H N_m)^{-1} N_m^T \underline{d}_u \qquad (3.3.28)$$

$$= - (N_m^T H N_m)^{-1} (N_m^T \underline{x}_u - \underline{b}_m)$$

$$\underline{d} = \underline{d}_u + H N_m \underline{\lambda}_k . \qquad (3.3.29)$$

If $\underline{d}_u$ (3.3.18) and $\underline{d} = \underline{x}_c - \underline{x}_k$ are substituted in (3.3.29), the projection (3.3.6) is obtained. Goldfarb's (1972) and Fletcher's (1971) algorithms solve (3.3.27) and

$$\nabla q(\underline{x}_k + \underline{d}) = \nabla q(\underline{x}_k) + G\underline{d} = N_m \underline{\lambda}_k . \qquad (3.3.30)$$

Hence

$$\underline{d} = - H \nabla q(\underline{x}_k) + H N_m \underline{\lambda}_k . \qquad (3.3.31)$$

Premultiplying (3.3.31) by N_m^T and applying (3.3.27) yields

$$N_m^T \underline{d} = \underline{o} = - N_m^T H \nabla q(\underline{x}_k) + N_m^T H N_m \underline{\lambda}_k .$$

Thus, the solution can be written as

$$\underline{\lambda}_k = (N_m^T H N_m)^{-1} N_m^T H \nabla q(\underline{x}_k) \qquad (3.3.32)$$

$$\underline{d} = -(H - H N_m(N_m^T H N_m)^{-1} N_m^T H) \nabla q(\underline{x}_k) \quad . \qquad (3.3.33)$$

As $\underline{d} = \underline{x}_c - \underline{x}_k$, the computation of (3.3.33) involves the evaluation of $\nabla q(\underline{x}_k)$ which is not required for (3.3.6). Instead, (3.3.6) requires the residual vector $N_m^T \underline{x}_u - \underline{b}_m$ which is part of the larger residual vector $N_{m_o}^T \underline{x}_u - \underline{b}_{m_o}$ that is evaluated only once. It should be noted that $\underline{\lambda}_k$ in (3.3.32) is identical to $\underline{\lambda}_c$ in (3.3.5) since

$$\begin{aligned} N_m^T H \nabla q(\underline{x}_k) &= N_m^T H(G(\underline{x}_k - \underline{x}_u)) \\ &= N_m^T \underline{x}_k - N_m^T \underline{x}_u \\ &= \underline{b}_m - N_m^T \underline{x}_u \quad . \end{aligned} \qquad (3.3.34)$$

For each computation of $\underline{\lambda}_k$ and $\underline{d}$, (3.3.32) - (3.3.33) require $2n^2 + mn$ arithmetic operations counting only multiplications and divisions and assuming $(N_m^T H N_m)^{-1} N_m^T H$ and $(H - HN_m(N_m^T HN_m)^{-1} N_m^T H)$ are given. The arithmetic operations required for computing $\underline{\lambda}_c$ and $\underline{x}_c$ with (3.3.5) - (3.3.6) is $m^2 + mn + n^2$ assuming $\underline{x}_u$, $(N_m^T H N_m)^{-1}$ and $N_m^T \underline{x}_u - \underline{b}_m$ are given. Since $m \le n$ computing (3.3.5) - (3.3.6) is at least as good or better than (3.3.32) - (3.3.33). In updating the respective operators used in these expressions when the active set is changed, (3.3.5) - (3.3.6) will be shown, in Section (3.5.2), to require at least n^2 less arithmetic operations than (3.3.32) - (3.3.33). This demonstrates the decisive superiority of (3.3.5) - (3.3.6) over (3.3.32) - (3.3.33).

Bartels et. al. (1970) use the Cholesky factors $\hat{L}\,\hat{D}\,\hat{L}^T$ of G and Householder transformations to construct the projection in (3.3.33).

Thus, (3.3.32) and (3.3.33) are replaced by (see Goldfarb (1975)),

$$\underline{\lambda}_k = L^{-T} Q_1^T R^{-T} \nabla q(\underline{x}_k) \cdot 1_1 \qquad (3.3.35)$$

$$\underline{d} = - R^{-1} Q_2 Q_2^T R^{-T} \nabla q(\underline{x}_k) , \qquad (3.3.36)$$

where

$$R^T = \hat{L} \hat{D}^{\frac{1}{2}} , \qquad (3.3.37)$$

$$\hat{L} \hat{D} \hat{L}^T = G ,$$

$$R^{-T} N_m = [Q_1 \vdots Q_2] \begin{bmatrix} L^T \\ \hline 0 \end{bmatrix} . \qquad (3.3.38)$$

If

$$\nabla q(\underline{x}_k) = G(\underline{x} - \underline{x}_u) = \hat{L} \hat{D} \hat{L}^T (\underline{x}_k - \underline{x}_u)$$

is substituted in (3.3.35) - (3.3.36), the simplified expressions

$$\underline{\lambda}_k = L^{-T} Q_1^T R(\underline{x}_k - \underline{x}_u) \qquad (3.3.39)$$

$$\underline{d} = - R^{-1} Q_2 Q_2^T R(\underline{x}_k - \underline{x}_u) \qquad (3.3.40)$$

are obtained that require n^2 less arithmetic operations than (3.3.35) - (3.3.36) since the gradient $\nabla q(\underline{x}_k) = G \underline{x}_k + \underline{a}$ is no longer computed. It should be noted that the quadratic programming method by Bartels et. al. (1970) involves the solution of a linear complimentarity problem rather than successive computations of $\underline{d}$ and $\underline{\lambda}$ for solving the inequality constrained problem (3.1.7).

3.3.2 Inequality Constraints and the Unconstrained Optimum

The solution of the inequality constrained quadratic programming problem (3.1.5) could be obtained in one step (3.3.5) - (3.3.6) if the set of the active constraints at the constrained solution $\underline{x}_c$ were known. Also, Theorem (3.1) could be invoked to interpret $\underline{x}_c$ as the projection of $\underline{x}_u$ onto the set of active constraints at $\underline{x}_c$.

If $\underline{x}_u$ is feasible, it is also optimal for the constrained problem. If not, for a convex objective function $f(\underline{x})$ and a general convex region R (1.1.22), $\underline{x}_c$ lies on the boundary of R. This follows directly from the property of the convex function $f(\underline{x})$ that for $\underline{x}_u$ such that

$$\nabla f(\underline{x}_u) = \underline{o} \quad , \qquad f(\underline{x}_u) \leq f(\underline{x}) \ , \ \forall \underline{x} \in E^n \qquad (3.3.41)$$

(Polak (1971), Theorem B. 2.7).

One further property of convex $f(\underline{x})$ and R is the relationship of the violated constraints at $\underline{x}_u$ and the active constraints at $\underline{x}_c$, the constrained optimum of $f(\underline{x})$ subject to R. This is stated in the following theorem.

Theorem (3.2)

For convex $f(\underline{x})$ and R (1.1.22), if $\underline{x}_u$ satisfying (3.3.41) violates any constraints (i.e. $\underline{x}_u \notin R$), at least one of these constraints is active at $\underline{x}_c$.

Proof

Premultiplying the optimality condition (1.1.24) by $(\underline{x}_c - \underline{x}_u)^T$ we get

$$< (\underline{x}_c - \underline{x}_u) , \nabla f(\underline{x}_c) > = < (\underline{x}_c - \underline{x}_u), \nabla \underline{g}(\underline{x}_c) \underline{\lambda}_c > .$$

The inequality

$$< \underline{x}_c - \underline{x}_u , \nabla f(\underline{x}_c) > \;\; \geq \;\; f(\underline{x}_c) - f(\underline{x}_u) \geq o$$

follows the convexity of $f(\underline{x})$ (Polak (1971) Theorem B. 2.4) and (3.3.41). Thus

$$< \underline{x}_c - \underline{x}_u , \nabla \underline{g}(\underline{x}_c) \underline{\lambda}_c > \;\; \geq \;\; o \; . \qquad (3.3.42)$$

The hyperplane

$$h^o = \{ \underline{x} \in E^n \mid < \underline{g}(\underline{x}_c), \underline{\lambda}_c > + < \underline{x} - \underline{x}_c, \nabla \underline{g}(\underline{x}_c) \underline{\lambda}_c > \; = \; o \} \qquad (3.3.43)$$

supports R at $\underline{x}_c$ (Luenberger (1969) pp. 131 - 137). Furthermore the complimentarity condition

$$< \underline{\lambda}_c , \underline{g}(\underline{x}_c) > \; = \; o \qquad (1.1.26)$$

may be used to simplify (3.3.43), hence

$$h^o = \{ \underline{x} \in E^n \mid < \underline{x} - \underline{x}_c , \nabla \underline{g}(\underline{x}_c) \underline{\lambda}_c > = o \} \qquad (3.3.44)$$

Note that since $\underline{\lambda}_c \geq \underline{o}$, the half space

$$h = \{ \underline{x} \in E^n \mid < \underline{x} - \underline{x}_c, \nabla \underline{g}(\underline{x}_c) \underline{\lambda}_c > \; \geq o \} \qquad (3.3.45)$$

contains the region R (1.1.22). For $\underline{x}_u \neq \underline{x}_c$ (3.3.42) becomes a strict inequality for $f(\underline{x})$ strictly convex. Thus, $\underline{x}_u$ is in the complement of the half space h and $\underline{x}_u \notin R$. As $\underline{\lambda}_c \geq \underline{o}$ for at least one i with the ith element of $\underline{\lambda}_c$, $\lambda_c^i > o$,

$$< \underline{x}_c - \underline{x}_u , \nabla \underline{g}_i(\underline{x}_c) > \; > \; o \qquad (3.3.46)$$

where $\nabla \underline{g}_i$ is the ith column of $\nabla \underline{g}$. Because of the convexity of R, (3.3.46) contains the constraint $g_i(\underline{x}) \geq o$. Thus $g_i(\underline{x}_u) < o$ for at least one i. □

Theorem (3.2) clearly applies to the quadratic programming problem (3.1.7). (For an alternative proof of this theorem see, e.g. Mazzoleni (1975).)

The unconstrained optimum of the quadratic function $q(\underline{x})$ has been used in a number of existing quadratic programming algorithms. Theil and Van de Panne (1960) have designed a method that searches for the set of active constraints at $\underline{x}_c$ by enumerating all possible combinations of the inequality constraints as equalities and projecting $\underline{x}_u$ in their intersection. Golub and Saunders (1970), Bartels et. al. (1970) use $\underline{x}_u$ to define a linear complimentarity problem which is solved to determine the active constraints at $\underline{x}_c$. Subsequently, the equality constrained problem (3.3.1) is computed to obtain $\underline{x}_c$.

As a simple example to illustrate Theorem (3.2) consider the problem (Beale (1967), Zoutendijk (1970))

$$\min \{ -4x_1 - 3x_2 - 2x_3 + x_1^2 + x_2^2 + \tfrac{1}{2}x_3^2 + x_1x_2 + x_1x_3 \mid$$

$$\mid -x_1 - x_2 - 2x_3 \geq -3, \; x_1 \geq 0, \; x_2 \geq 0, \; x_3 \geq 0 \} .$$

Computing

$$\underline{x}_u = -H\,\underline{a} = \begin{bmatrix} 2 & -1 & -2 \\ -1 & 1 & 1 \\ -2 & 1 & 3 \end{bmatrix} \begin{bmatrix} -4 \\ -3 \\ -2 \end{bmatrix} = \begin{bmatrix} 1 \\ 1 \\ 1 \end{bmatrix} ,$$

the only constraint violated at $\underline{x}_u$ is the first. Taking

$$N_1^T = [\,-1 \;\; -1 \;\; -2\,] \;, \quad b_1 = -3 \;,$$

as this is the only violated constraint, the projection of $\underline{x}_u$ on it is computed. By Theorems (3.1) - (3.2), if the projection does not violate any remaining constraints, it will also be the constrained optimum. Hence, using (3.3.6), the projection

$$\underline{x} = \underline{x}_u - H\, N_1 (N_1^T H\, N_1)^{-1} (N_1^T \underline{x}_u - b_1)$$

$$= \begin{bmatrix} 1 \\ 1 \\ 1 \end{bmatrix} - \begin{bmatrix} 3 \\ -2 \\ -5 \end{bmatrix} [\,9\,]^{-1} [\,-1\,]$$

$$= \begin{bmatrix} 4/3 \\ 7/9 \\ 4/9 \end{bmatrix}$$

is computed. As $\underline{x}$ is feasible, $\underline{x}_c = \underline{x}$. Clearly, this example does not illustrate the problems that may arise if $\underline{x}$ is not feasible. This, and related problems will be discussed in the next section.

Alternatively, dropping some of the active constraints and projecting $\underline{x}_u$ on the remaining ones may reduce the objective function further. Consider the feasible descent direction, $\underline{d}$, at $\underline{x}_k$ where

$$\underline{x}_k = \underline{x}_u - H\, N_m (N_m^T H\, N_m)^{-1} (N_m^T \underline{x}_u - \underline{b}_m) \tag{3.3.47}$$

hence. $N_m \underline{x}_k = \underline{b}_m$ and $N_m^T \underline{d} \geq \underline{o}$. Clearly,

$$N_m \underline{\lambda}_k = \nabla q(\underline{x}_k) = G(\underline{x}_k - \underline{x}_u) \tag{3.3.48}$$

holds and premultiplying (3.3.48) by $\underline{d}$

$$< \underline{d},\, N_m \underline{\lambda}_k > \; = \; < \underline{d},\, G(\underline{x}_k - \underline{x}_u) > \tag{3.3.49}$$

is obtained. Thus, $\underline{d}$ is a descent direction only if (3.3.49) is negative. Since $N_m^T \underline{d} \geq \underline{o}$, this can only occur if some elements of $\underline{\lambda}_k$ are negative. On the other hand, if all elements of $\underline{\lambda}_m$ are positive, (3.3.49) implies that there is no feasible descent direction at $\underline{x}_k$.

If there exists an element of $\underline{\lambda}$, say $\lambda^i < o$, then the projection of $\underline{x}_u$ on the remaining $m - 1$ constraints is

$$\underline{x}_{k+1} = \underline{x}_u - H N_{m-1} (N_{m-1}^T H N_{m-1})^{-1} (N_{m-1}^T \underline{x}_u - \underline{b}_{m-1}) \tag{3.3.50}$$

where N_{m-1} is obtained by deleting the ith column of N_m and $\underline{b}_{m-1}$ is obtained by deleting the ith element of $\underline{b}_m$. To show that

$$\underline{d} = \underline{x}_{k+1} - \underline{x}_k$$

is a descent direction consider (3.3.49)

$$\begin{aligned} < \nabla q(\underline{x}_k), \underline{d} > &= < \underline{d} , G(\underline{x}_k - \underline{x}_u) > \\ &= < \underline{x}_{k+1} - \underline{x}_k , G(\underline{x}_k - \underline{x}_u) > \\ &= < \underline{x}_{k+1} - \underline{x}_k , G(\underline{x}_k - \underline{x}_{k+1} + \underline{x}_{k+1} - \underline{x}_u) > \\ &< < \underline{x}_{k+1} - \underline{x}_k , G(\underline{x}_{k+1} - \underline{x}_u) > \\ &\leq o \end{aligned}$$

where the last inequality follows from the fact that $\underline{x}_{k+1}$ is a projection of $\underline{x}_u$ (see Lemma (4.2)). Thus, for $\underline{x}_{k+1} \neq \underline{x}_k$,

$$< \nabla q(\underline{x}_k), \underline{d} > = < G(\underline{x}_k - \underline{x}_u), \underline{d} > < o . \tag{3.3.51}$$

Since $N_{m-1} \underline{d} = \underline{o}$ and $\lambda^i < o$, it follows from (3.3.51) and (3.3.49) that

$$< \underline{n}_i , \underline{d} > > o . \tag{3.3.52}$$

Hence, dropping constraint i, $\lambda^i < o$, and projecting $\underline{x}_u$ onto the remaining constraints, yields $\underline{x}_{k+1}$ which both reduces the objective function and satisfies constraint i. It also follows from above that, in general, $\underline{\lambda}$ computed satisfying (3.3.48) correctly identifies the constraint to be dropped. Any descent direction $\underline{d}$ computed in the intersection of the remaining constraints is also feasible with respect to constraint i. Only one of the methods discussed by Gill and Murray (1977) was shown to have this property.

In view of (3.3.49), $\underline{d}$ can be chosen to be orthogonal to all those active constraints with nonnegative Lagrange multipliers. This can be done by dropping all active constraints with negative multipliers from the active set and computing $\underline{d} = \underline{x}_{k+1} - \underline{x}_k$ such that $\underline{x}_{k+1}$ is the projection of $\underline{x}_u$ onto the constraint hyperplanes remaining in the active set. It follows from (3.3.49) that in this case $\underline{d}$ is a descent direction. Goldfarb (1972) discusses an algorithm that can drop more than one constraint at a time. Gill and Murray (1978) recognise that a descent direction is obtained by dropping all the constraints with negative multipliers but choose, in practise, to drop one constraint at a time and recalculate $\underline{d}$ in the intersection of the remaining active constraints. Fletcher (1971) also drops one constraint at a time.

3.4 THE ALGORITHM

In this section a positive definite quadratic programming algorithm and related problems are discussed. The algorithm is described in Section (3.4.1). The extension of the algorithm to cases when the second derivative matrix of the objective function is indefinite is discussed in Section (3.4.2). The extension of the algorithm to nonlinear constraints is discussed in Section (3.4.3).

3.4.1 A Positive Definite Quadratic Programming Algorithm

The algorithm below does not handle equality constraints explicitly. For such constraints, the simple rule that requires them to remain in the active set can be appended to the algorithm.

The following steps describe an iteration of the algorithm.

Each iteration defines a projection of $\underline{x}_u$ onto a different active set. Thus, a fixed set of active constraints characterises an iteration.

Algorithm (3.4.1)

Step 0: Given H, $\underline{a}$, compute $\underline{x}_u = - H \underline{a}$. Compute

$$\underline{\theta}_{m_o}(\underline{x}_u) = N_{m_o}^T \underline{x}_u - \underline{b}_{m_o} \qquad (3.4.2)$$

If $\underline{\theta}_{m_o} \geq \underline{o}$ the solution is at $\underline{x}_u$; the unconstrained optimum: stop . Otherwise, given a feasible point $\underline{x}_o$ (see Chapter 2), compute

$$\underline{\theta}_{m_o}(\underline{x}_o) = N_{m_o}^T \underline{x}_o - \underline{b}_{m_o} . \qquad (3.4.3)$$

The elements of $\underline{\theta}_{m_o}(\underline{x}_o)$ that are strictly zero correspond to the active constraints at $\underline{x}_o$. The strictly positive elements of $\underline{\theta}_{m_o}(\underline{x}_o)$ correspond to the strictly satisfied constraints at $\underline{x}_o$. Set $k = o$.

Step 1: Construct the index of active constraints at $\underline{x}_k$, $I(\underline{x}_k)$.

Construct the vector $\underline{\theta}_m(\underline{x}_u)$ from the elements of $\underline{\theta}_{m_o}(\underline{x}_u)$ corresponding to the active constraints at $\underline{x}_k$.

Construct N_m as the matrix whose columns are the normals of the active constraints. Similarly, set up vector $\underline{b}_m$.

Construct the matrix $(N_m^T H N_m)^{-1}$.

Construct the set of strictly satisfied constraints, $S(\underline{x}_k)$ at $\underline{x}_k$.

Construct the vector $\underline{\theta}_s(\underline{x}_k)$ as the residuals of the constraints in $S(\underline{x}_k)$. Thus, the jth element of $\underline{\theta}_s(\underline{x}_k)$ is given by

$$\theta_{s_j}(\underline{x}_k) = \langle \underline{n}_j, \underline{x}_k \rangle - b_j \, , \quad j \in S(\underline{x}_k) \, . \tag{3.4.4}$$

Step 2: Compute the vector of multipliers

$$\underline{\lambda}_{k+1} = -(N_m \, H \, N_m)^{-1} \, \underline{\theta}_m(\underline{x}_u) \tag{3.4.5}$$

and the projection of $\underline{x}_u$ onto the active constraints

$$\underline{x}_{k+1} = \underline{x}_u + H \, N_m \, \underline{\lambda}_{k+1} \tag{3.4.6}$$

Step 3: Test for constraint violations at $\underline{x}_{k+1}$ among the strictly satisfied constraints at $\underline{x}_k$. Hence, compute

$$\theta_s^i(\underline{x}_{k+1}) = \langle \underline{n}_j, \underline{x}_{k+1} \rangle - b_j \, . \tag{3.4.7}$$

If $\underline{\theta}_s(\underline{x}_{k+1}) \geq \underline{o}$, $\underline{x}_{k+1}$ is feasible; go to Step 4. Otherwise determine the maximum step size in going from $\underline{x}_k$ to $\underline{x}_{k+1}$ such that no constraint is violated, i.e.

$$\tau = \min_{j \in S(\underline{x}_k)} \left\{ \frac{-\theta_s^j(\underline{x}_k)}{\theta_s^j(\underline{x}_{k+1}) - \theta_s^j(\underline{x}_k)} \right\} . \tag{3.4.8}$$

Select the constraint(s) corresponding to τ to be added to the active set. If there are more than one constraints to be added, the linear dependence, on the active constraint normals, of those added after the first one must be tested. This is discussed further in Section (3.4.3).

Set

$$\underline{x}_{k+1} = \underline{x}_k + \tau(\underline{x}_{k+1} - \underline{x}_k) \tag{3.4.9}$$

$$k = k + 1$$

and go to Step 1.

Step 4: Determine the minimum element of the multiplier vector $\underline{\lambda}_{k+1}$,

$$\lambda_{min} = \min_{i=1,2,\ldots,m} \{\lambda_{k+1}^{i}\} .$$

If $\lambda_{min} \geq 0$, $\underline{x}_{k+1}$ is the optimal solution; stop. Otherwise, drop the constraint, i, corresponding to λ_{min}. Set $k = k+1$, go to Step 1.

In this algorithm H was assumed to be given. This is a reasonable assumption when the algorithm is used with the nonlinear programming methods discussed in Chapter 4. However, in general, G is more readily available. In this case, direct inversion can always be avoided by formulating the problems as solutions of linear equations (see, e.g. Goldfarb (1975)).

Algorithm (3.4.1) does not consider the matrix decompositions discussed in Section (3.3.1) for simplicity. Incorporating these to the algorithm can be done by suitably augmenting Step 1 above.

The origin of (3.4.8) lies in the distance from $\underline{x}_k$ to the hyperplane $\langle \underline{n}_j, \underline{x} \rangle - b_j = 0$ along the direction $\underline{x}_{k+1} - \underline{x}_k$. This distance is given by

$$- \frac{\langle \underline{n}_j, \underline{x}_k \rangle - b_j}{\langle \underline{n}_j, \underline{x}_{k+1} - \underline{x}_k \rangle}$$

which, in view of (3.4.4) and (3.4.7), may be expressed as (3.4.8) .

3.4.2 The Positive Semi-Definite and Indefinite Cases

The main objective of the algorithm presented in the previous section is to solve the positive definite quadratic programming problem. It may also be modified to solve problems with objective functions of the type

$$< \underline{x} - \underline{x}_u , G (\underline{x} - \underline{x}_u) > \qquad (3.4.10)$$

where $\underline{x}_u$ is given and G need not be strictly positive definite. In this general case the eigenvalues of G may be positive, zero or negative.

When G is semi-definite, and rank (G) < n, some eigenvalues of G are positive and some are zero. The zero eigenvalues imply that the quadratic function is an infinite valley (Gill and Murray (1974)) which has a horizontal floor if an eigenvector, corresponding to a zero eigenvalue, is orthogonal to $G\underline{x}_u$ but which slopes down to $-\infty$ otherwise. A bounded weak minimum to (3.4.10) exists only if $G\underline{x}_u$ is orthogonal to all eigenvectors, $\underline{v}$, of G corresponding to zero eigenvalues such that

$$G \underline{v} = \underline{o} .$$

Clearly, this minimum is $\underline{x}_u$. In the constrained case, a bounded minimum may exist even when a bounded unconstrained minimum does not. Such a bounded constrained minimum is guaranteed as long as the constraints bound the descent along the infinite valley mentioned above.

For the positive semidefinite case Fletcher (1971) suggests the computation of (3.3.32) and (3.3.33) only for bases which have a feasible point that is feasible or near feasible. This idea is closely associated with the positive definiteness condition on the projected Hessian, $(Z^T G Z)$, in the intersection of the active constraint hyperplanes. The positive definiteness of $(Z^T G Z)$ ensures the existence of a constrained minimum in this intersection (Gill and Murray (1978)), Gill and Murray (1974) give an example in which the Hessian is singular, but the projected Hessian, in the intersection of the constraints active at the solution, is not.

When G is indefinite, (3.4.10) does not possess a bounded minimum. However, in the constrained case an indefinite G may result in stationary points that can be local minima, maxima or saddle points. If at a stationary point the projected Hessian is positive definite, the quadratic function may not be reduced further. However, if the projected Hessian is indefinite at a stationary point a descent direction, say $\underline{d}$, known as a direction of negative curvature, may be computed such that $<\underline{d}, G\,\underline{d}> \; < o$. If a step along a direction of negative curvature is taken, the objective function must be reduced. Gill and Murray (1978) discuss computational procedures for such directions.

3.4.3 Extensions to Nonlinear Constraints

The extension of Algorithm (3.4.1) to nonlinear constraints is possible since Theorem (3.1) still holds when the constraints are nonlinear. Consider the first order optimality conditions for the nonlinear equality constrained problem

$$\min \; \{\tfrac{1}{2} \|\underline{x} - \underline{x}_u\|_G^2 \mid \underline{g}(\underline{x}) = \underline{o}\} \tag{3.4.11}$$

where $\underline{g}$ is an m-dimensional function. The first order conditions for (3.4.11) are

$$G(\underline{x}_c - \underline{x}_u) - N_m(\underline{x}_c)\,\underline{\lambda}_c = \underline{o} \qquad (3.4.12,a)$$

$$\underline{g}(\underline{x}_c) = \underline{o} \qquad (3.\ 4.12,b)$$

where $n \times m$ matrix $N_m(\underline{x}_c)$ is defined in the same way as $N_{m_o}(\underline{x}_k)$ in (1.2.6). As in the linear case, the columns of $N_m(\underline{x}_k)$ are the gradients of the constraints evaluated at $\underline{x}_k$. The columns of N_m will be assumed to be linearly independent. Substituting the linear approximation (1.2.5) to $\underline{g}(\underline{x})$ at $\underline{x}_k$ in (3.4.12), an iterative scheme based on

$$G(\underline{x}_{k+1} - \underline{x}_u) - N_m(\underline{x}_k)\,\underline{\lambda}_{k+1} = \underline{o} \qquad (3.4.13,a)$$

$$\underline{g}(\underline{x}_k) + N_m^T(\underline{x}_k)\,(\underline{x}_{k+1} - \underline{x}_k) = \underline{o} \qquad (3.4.13,b)$$

may be constructed. Thus $\underline{x}_{k+1}$, $\underline{\lambda}_{k+1}$ are intended to satisfy (3.4.12). From (3.4.13,a)

$$\underline{x}_{k+1} = \underline{x}_u + H\,N_m(\underline{x}_k)\,\underline{\lambda}_{k+1} \qquad (3.4.14)$$

may be obtained. Substituting this in (3.4.13,b)

$$\underline{o} = \underline{g}(\underline{x}_k) + N_m^T(\underline{x}_k)(\underline{x}_{k+1} - \underline{x}_u + \underline{x}_u - \underline{x}_k)$$

$$= \underline{g}(\underline{x}_k) + N_m^T(\underline{x}_k)(H\,N_m(\underline{x}_k)\,\underline{\lambda}_{k+1} + \underline{x}_u - \underline{x}_k)$$

and solving for $\underline{\lambda}_{k+1}$ yields

$$\underline{\lambda}_{k+1} = (N_m^T(\underline{x}_k)\,H\,N_m(\underline{x}_k))^{-1}\,(N_m^T(\underline{x}_k)(\underline{x}_k - \underline{x}_u) - g(\underline{x}_k))\ . \qquad (3.4.15)$$

Thus an algorithm for nonlinear constraints may be based on (3.4.14) and (3.4.15). Inequality constraints may be resolved either by introducing slack variables to transform these to equality constraints or by using an active set strategy to prevent $\underline{x}_{k+1}$ from violating already satisfied constraints.

Since

$$\underline{g}(\underline{x}_u) \cong \underline{g}(\underline{x}_k) + N_m^T(x_k)(\underline{x}_u - \underline{x}_k) \tag{3.4.16}$$

is the linear approximation of $\underline{g}$ at $\underline{x}_k$ evaluated at $\underline{x}_u$, an alternative to (3.4.15) is

$$\underline{\lambda}_{k+1} = -(N_m^T(\underline{x}_k)\ H\ N_m(\underline{x}_k))^{-1}\ \underline{g}(\underline{x}_u)\ , \tag{3.4.17}$$

whenever (3.4.16) holds.

The argument against the above approach to nonlinear constrained problems is that it does not make proper use of second derivative information of the constraints. Thus information about the curvature of the constraints is not being used.

An alternative approach that also compensates for the curvature of the constraints involves quadratic approximations to the constraints. Thus, each constraint (i.e. each element of $\underline{g}$) may be approximated by

$$g_i(\underline{x}) = g_i(\underline{x}_k) + <\nabla g_i(\underline{x}_k), \underline{x} - \underline{x}_k> + \tfrac{1}{2}<\underline{x} - \underline{x}_k, G_k^i(\underline{x} - \underline{x}_k)> \tag{3.4.18}$$

where $\nabla g_i(\underline{x}_k)$ is the ith column of $N_m(\underline{x}_k)$ and G_k^i is the $n \times n$ dimensional second derivative matrix of constraint i, evaluated at $\underline{x}_k$. For $\underline{x}_{k+1}$ and $\underline{\lambda}_{k+1}$ intended to satisfy (3.4.12), the optimality condition (3.4.12,a) may be evaluated using the second order expansion (3.4.18) to yield

$$G(\underline{x}_{k+1} - \underline{x}_u) - N_m(\underline{x}_k)\,\underline{\lambda}_{k+1} - \sum_{i=1}^{m} \lambda_{k+1}^{i}\, G_k^{i}(\underline{x}_{k+1} - \underline{x}_k) = \underline{o} \tag{3.4.19}$$

where λ_{k+1}^{i} is the ith element of $\underline{\lambda}_{k+1}$. Rewriting (3.4.19)

$$(G - \sum_{i=1}^{m} \lambda_{k+1}^{i}\, G_k^{i})(\underline{x}_{k+1} - \underline{x}_k) + G(\underline{x}_k - \underline{x}_u) - N_m(x_k)\underline{\lambda}_{k+1} = \underline{o}$$

and setting

$$G_k^L = (G - \sum_{i=1}^{m} \lambda_{k+1}^{i}\, G_k^{i}) \tag{3.4.20}$$

an expression for $\underline{x}_{k+1}$ can be obtained

$$\underline{x}_{k+1} = \underline{x}_k - H_k^L(G\,(\underline{x}_k - \underline{x}_u) - N_m(\underline{x}_k)\underline{\lambda}_{k+1}). \tag{3.4.21}$$

where $H_k^L = (G_k^L)^{-1}$. At this stage G_k is assumed to be positive definite. Thus, H_k^L exists. Substituting (3.4.21) in (3.4.13,b) yields an expression for $\underline{\lambda}_{k+1}$

$$\underline{o} = \underline{g}(\underline{x}_k) \qquad + \qquad N_m^T(\underline{x}_k)(H_k^L\, N_m(\underline{x}_k)\underline{\lambda}_{k+1} - H_k^L\, G(\underline{x}_k - \underline{x}_u))$$

$$\underline{\lambda}_{k+1} = -\,(N_m^T(\underline{x}_k)H_k^L N_m^T(\underline{x}_k))^{-1}(\,\underline{g}(\underline{x}_k) - N_m^T(\underline{x}_k)H_k^L\, G(\underline{x}_k - \underline{x}_u))\,. \tag{3.4.22}$$

Thus, (3.4.21) and (3.4.22) may form the basis of a method that makes use of the curvature of the constraints. If it is assumed that

$$G(\underline{x}_k - \underline{x}_u) - N_m(\underline{x}_k)\underline{\lambda}_k = \underline{o}$$

holds, (3.4.21) and (3.4.22) may be simplified to

$$\underline{x}_{k+1} = \underline{x}_k + H_k^L\, N_m(\underline{x}_k)(\underline{\lambda}_{k+1} - \underline{\lambda}_k) \tag{3.4.23}$$

and

$$\underline{\lambda}_{k+1} - \underline{\lambda}_k = -\,(N_m^T(\underline{x}_k)\, H_k^L\, N_m(\underline{x}_k))^{-1}\, \underline{g}(\underline{x}_k)\,. \tag{3.4.24}$$

The matrix G_k^L is not known until $\underline{\lambda}_{k+1}$ is computed and thus has to be replaced by

$$G_k^L = (H^L)^{-1} = (G - \sum_{i=1}^{m} \lambda_k^i \, G_k^i) \qquad (3.4.25)$$

Hence all subsequent use of G_k^L or H_k^L will refer to (3.4.25).

An alternative approach is to construct an approximation to the Lagrangian function of (3.4.11) at $\underline{x}_k$ and $\underline{\lambda}_k$,

$$\ell(\underline{x},\underline{\lambda}_k) = \tfrac{1}{2} <\underline{x} - \underline{x}_u, G(\underline{x} - \underline{x}_u)> - \sum_{i=1}^{m} \lambda_k^i (\tfrac{1}{2} <\underline{x} - \underline{x}_k, G_k^i(\underline{x} - \underline{x}_k)>) \qquad (3.4.26)$$

and minimise it subject to the linearised constraints (3.4.13,b). This also explains the reason for omitting from (4.2.26) the linear terms $- <\lambda_k, N_m^T(\underline{x}_k)(\underline{x} - \underline{x}_k)>$ of the constraints. Hence, the problem becomes

$$\min \{ \ell(\underline{x},\underline{\lambda}_k) \mid N_m^T(\underline{x}_k)(\underline{x} - \underline{x}_k) + \underline{g}(\underline{x}_k) = \underline{o} \} . \qquad (3.4.27)$$

The Lagrangian of (3.4.27) may be written as

$$L(\underline{x},\underline{\lambda}) = \ell(\underline{x},\underline{\lambda}_k) - < N_m^T(\underline{x}_k)(\underline{x} - \underline{x}_k) + \underline{g}(\underline{x}_k), \underline{\lambda} > . \qquad (3.4.28)$$

The first order optimality conditions for (3.4.27) can be written using (3.4.28)

$$\begin{aligned} \nabla_x L(\underline{x},\underline{\lambda}) &= G(\underline{x} - \underline{x}_u) - \sum_{i=1}^{m} \lambda_k^i \, G_k^i(\underline{x} - \underline{x}_k) - N_m(\underline{x}_k)\underline{\lambda} \\ &= (G - \sum_{i=1}^{m} \lambda_k^i \, G_k^i)(\underline{x} - \underline{x}_k) - N_m(\underline{x}_k)\underline{\lambda} + G(\underline{x}_k - \underline{x}_u) = \underline{o} , \end{aligned} \qquad (3.4.29,a)$$

$$N_m^T(\underline{x}_k)(\underline{x} - \underline{x}_k) + \underline{g}(\underline{x}_k) = \underline{o} . \qquad (3.4.29,b)$$

Let $\underline{x}_{k+1}, \underline{\lambda}_{k+1}$ denote the values of $\underline{x},\underline{\lambda}$ that solve (3.4.27). These can be obtained from (3.4.29) to be

$$\underline{x}_{k+1} = \underline{x}_k - H_k^L (G (\underline{x}_k - \underline{x}_u) - N_m(\underline{x}_k)\underline{\lambda}_{k+1}) \qquad (3.4.30,a)$$

$$\underline{\lambda}_{k+1} = -(N_m^T(\underline{x}_k)\, H_k^L\, N_m(\underline{x}_k))^{-1}\, (\underline{g}(\underline{x}_k) - N_m^T(\underline{x}_k)\, H_k^L\, G(\underline{x}_k - \underline{x}_u)). \tag{3.4.30,b}$$

For inequality constraints, (3.4.26) is defined for all constraints. Thus, m is replaced by m_o, the number of inequality constraints, and $\underline{\lambda}$ becomes m_o-dimensional. This function is minimised subject to all the linearised constraints (1.2.5). Thus the problem can be written as

$$\min \{\ell(\underline{x},\underline{\lambda}_k) \mid N_{m_o}^T(\underline{x}_k)(\underline{x} - \underline{x}_k) + \underline{g}(\underline{x}_k) \geq \underline{o}\} . \tag{3.4.31}$$

This is basically an application of the Newton-type approach, developed by Wilson (1963) for nonlinear programming, to quadratic objective functions. Wilson's algorithm is discussed in Section (1.2.1).

As the unconstrained optimum of $\ell(\underline{x},\underline{\lambda}_k)$ is given by

$$\underline{x}_u = \underline{x}_k - H_k^L(G(\underline{x}_k - \underline{x}_u) \tag{3.4.32}$$

an equivalent formulation , in view of Theorem (3.1), may be written as

$$\min \{ \tfrac{1}{2}\| \underline{x} - \underline{x}_u \|^2_{G_k^L} \mid N_{m_o}^T(\underline{x}_k)(\underline{x} - \underline{x}_k) + \underline{g}(\underline{x}_k) \geq \underline{o} \} .$$

When G_k^c is positive semi definite or indefinite, the discussion in the previous section applies to problems (3.4.27) and (3.4.31).

Evaluating the second derivative for the constraints imposes a substantial computational requirement. This can be circumvented by approximating

$$-\sum_{i=1}^{m_o} \lambda_k^i\, G_k^i \tag{3.4.33}$$

with a matrix that is updated at $\underline{x}_k$, k=1,2,... with a variable metric formula. The matrices G^i of the constraints are negative semidefinite everywhere if the feasible region (1.1.22) described by the constraints is convex. This is implied by the convexity of the feasible region which requires the constraint functions to be concave (Fiacco and McCormick (1968)). However, since the above computations depend only G_k^L being positive definite, neither $-\sum_{i=1}^{m_o} \lambda_k^i G_k^i$ nor its approximation need to be positive definite.

To derive the necessary information used in variable metric updates to second derivative matrices or their inverses (see Section (4.1.2)) consider the second order approximation about $\underline{x}_k$ of the optimality condition

$$G(\underline{x}_c - \underline{x}_u) - N_{m_o}(\underline{x}_c)\,\underline{\lambda}_c = \underline{o} ,$$

evaluated at $\underline{x}_{k+1}$, $\underline{\lambda}_{k+1}$,

$$G(\underline{x}_{k+1} - \underline{x}_u) - N_{m_o}(\underline{x}_{k+1})\,\underline{\lambda}_{k+1}$$

$$= G(\underline{x}_{k+1} - \underline{x}_u) - N_{m_o}(\underline{x}_k)\underline{\lambda}_{k+1}$$

$$- \sum_{i=1}^{m_o} \lambda_{k+1}^i \; G_{k+1}^i \; (\underline{x}_{k+1} - \underline{x}_k) + O(\| \underline{x}_{k+1} - \underline{x}_k \|^2) \; . \quad (3.4.34)$$

Rearranging (3.4.34)

$$(N_{m_o}(\underline{x}_{k+1}) - N_{m_o}(\underline{x}_k))\underline{\lambda}_{k+1} = (\sum_{i=1}^{m_o} \lambda_{k+1}^i \; G_{k+1}^j)(\underline{x}_{k+1} - \underline{x}_k) + (\|\underline{x}_{k+1} - \underline{x}_k\|^2)$$

and ignoring the $O(\|\underline{x}_{k+1} - \underline{x}_k\|^2)$ terms yields

$$\underline{\gamma}^L = (\sum_{j=1}^{m_o} \lambda_{k+1}^j \; G_{k+1}^j)\underline{\delta} \qquad (3.4.35)$$

where

$$\underline{\gamma}^L = (N_{m_o}(\underline{x}_{k+1}) - N_{m_o}(\underline{x}_k))\underline{\lambda}_{k+1} \qquad (3.4.36)$$

$$\underline{\delta} = \underline{x}_{k+1} - \underline{x}_k \, . \tag{3.4.37}$$

Hence the matrix (3.4.33) is required to satisfy (3.4.35) at every iteration. The updating of (3.4.33) may be done by a suitable variable metric formula. As an example the rank one formula, that has no special requirements on the size of the step $\underline{\delta}$, is given by

$$- \sum_{j=1}^{m_o} \lambda_{k+1}^j \, G_{k+1}^j = - \sum_{j=o}^{m_o} \lambda_{k+1}^j \, G_k^j + \frac{\underline{\Gamma} \, \underline{\Gamma}^T}{\langle \underline{\Gamma}, \underline{\delta} \rangle} \tag{3.4.38}$$

where

$$\underline{\Gamma} = \underline{\gamma}^L - \Big(- \sum_{j=1}^{m_o} \lambda_k^j \, G_k^j\Big)\underline{\delta} \, .$$

The optimisation of a quadratic function subject to nonlinear constraints can be seen as a special case of optimisation of a general nonlinear objective function subject to nonlinear constraints. The convergence of methods solving the quadratic programming subproblem (3.4.31) and updating G_k^L with a variable metric formula, when the objective function is nonlinear, is discussed by Han (1976), Powell (1977), Garcia-Palomares and Mangasarian (1974).

The quadratic nature of the objective function leads to the simplification that only the contribution of the constraints needs updating in G^L. However, when solving the system of equations for $\underline{x}_{k+1}$, e.g.,

$$G_k^L \, (\underline{x}_{k+1} - \underline{x}_k) = G(\underline{x}_k - \underline{x}_u) + N_{m_o}(\underline{x}_k) \, \underline{\lambda}_{k+1}$$

an efficient way of inverting G_k^L is required. A variable metric update to H_k^L may alter the exactly known curvature of the quadratic objective function because of rounding-off errors. In practice, a variable metric approach may still be utilised and the resulting algorithm will be similar to those described by Han (1976) and

Powell (1977). An alternative approach is based on inverting G_k^L using the Cholesky factors of G_k^L. Assume that a rank two update satisfying (3.4.35) is given by

$$- \sum_{j=1}^{m_o} \lambda_{k+1}^j G_{k+1}^j = - \sum_{j=1}^{m_o} \lambda_k^j G_k^j + \alpha \underline{z}\, \underline{z}^T + \beta \underline{w}\, \underline{w}^T \tag{3.4.39}$$

where the vectors $\underline{z}$,$\underline{w}$ and scalars α and β depend on the chosen update. The rank one update is a special case of (3.4.39) with either α or $\beta = o$. Since G is positive definite and $- \sum_{j=1}^{m_o} \lambda_{k+1}^j G_{k+1}^j$ is assumed to be positive definite, the Cholesky factors

$$G_k^L = (G - \sum_{j=1}^{m_o} \lambda_k^j G_k^j) = L\, D\, L^T \tag{3.4.40}$$

are computed. Using (3.4.39) the expression

$$G_{k+1}^L = (G - \sum_{j=1}^{m_o} \lambda_{k+1}^j G_{k+1}^j) = L\, D\, L^T + \alpha \underline{z}\, \underline{z}^T + \beta \underline{w}\, \underline{w}^T$$

$$= L(D + \alpha \underline{p}\, \underline{p}^T + \beta \underline{s}\, \underline{s}^T)\, L^T \tag{3.4.41}$$

is obtained where

$$L\, \underline{p} = \underline{z} \qquad L\, \underline{s} = \underline{w}\ . \tag{3.4.42}$$

To obtain the Cholesky factors of the term in brackets in (3.4.41), first the Cholesky factors of $(D + \alpha \underline{p}\, \underline{p}^T)$ are computed and given these factors, the Cholesky factors of $((D + \alpha \underline{p}\, \underline{p}^T) + \beta \underline{s}\, \underline{s}^T)$ are computed, as suggested by Gill and Murray (1972), to yield

$$(D + \alpha \underline{p}\, \underline{p}^T + \beta \underline{s}\, \underline{s}^T) = \tilde{L}\, \tilde{D}\, \tilde{L}^T .$$

The Cholesky factors of G_{k+1}^L can be obtained using

$$G_{k+1}^L = L\, \tilde{L}\, \tilde{D}\, \tilde{L}^T L^T = \bar{L}\, \bar{D}\, \bar{L}^T \tag{3.4.43}$$

(see, Gill, Golub, Murray, Saunders (1974)), where

$$\bar{L} = L\, \tilde{L} \qquad \bar{D} = \tilde{D}\ .$$

Thus $G_{k+1}^{L} = \bar{L}\, D\, \bar{L}^{T}$ is used to compute the next point of the algorithm. Other approaches to computing and updating Cholesky factors of G_{k+1}^{L} are discussed by Goldfarb (1976).

3.5 RECURRENCE RELATIONS

In this section a number of recurrence relations will be discussed for updating the operators introduced in Section (3.3.1). The operators are updated as inequality constraints are added to or dropped from the active set. These relations will be used to illustrate various properties of the quadratic programming algorithm.

In Section (3.5.1) methods for updating

$$\bar{N}_m = (N_m^T H N_m)^{-1} N_m^T \qquad (3.5.1,a)$$

and

$$N_m^* = (N_m^T H N_m)^{-1} N_m^T H \qquad (3.5.1,b)$$

when a column is added to or removed from N_m are discussed. This operator may be used for computing $\underline{x}_{k+1}$ with (3.4.6). In Section (3.5.2) updates to $(N_m^T H N_m)^{-1}$ are described. Finally, the detection of linear dependence on N_m, of a column to be added to N_m is discussed in Section (3.5.3).

Updates of matrices such as $Z, (Z^T G Z)^{-1}$ and others mentioned in Section (3.5.1) are discussed by Gill and Murray (1978) and Goldfarb (1975) and will not be discussed further.

The discussion in Sections (3.5.1) and (3.5.2) on updating $\bar{N}_m$, N_m^* and $(N_m^T H N_m)^{-1}$ when a constraint normal is to be removed from N_m is based on the assumption that this normal corresponds to the last column in N_m and hence to the last rows of N_m^*, $\bar{N}_m$ and the last row and column of $(N_m^T H N_m)^{-1}$. This simplifying assumption can be relaxed. In case the constraint whose normal is in the ith column of N_m has to be dropped, N_m can be post multiplied with a matrix that exchanges the ith column of N_m with its last column. This matrix is the $m \times m$ identity matrix with its ith and last columns exchanged. If this matrix is denoted by $\tilde{I}$ then $\tilde{I}^T = \tilde{I}$ and $\tilde{I}\,\tilde{I} = I$. Thus,

$$\tilde{I}\,\bar{N}_m = (\tilde{I}\,N_m^T H N_m \tilde{I})^{-1} \tilde{I}\,N_m^T = \tilde{I}(N_m^T H N_m)^{-1} N_m^T$$

$$\tilde{I}\,N_m^* = (\tilde{I}\,N_m^T H N_m \tilde{I})^{-1} \tilde{I}\,N_m^T H = \tilde{I}(N_m^T H N_m)^{-1} N_m^T H$$

and hence exchanging the ith row of N_m with its last and then dropping the last row is equivalent to dropping the ith column of N_m. The same applies if the ith row and column of $(N_m^T H N_m)^{-1}$ are exchanged with its last row and column.

3.5.1 Updating N^* and Related Operators

When at $\underline{x}_{k+1}$,

$$\underline{x}_{k+1} = \underline{x}_u - N_m^{*T} (N_m^T \underline{x}_u - \underline{b}_m), \tag{3.5.2}$$

a constraint that was previously satisfied as a strict inequality, becomes active, i.e.

$$\langle \underline{n}, \underline{x}_{k+1} \rangle - b = 0, \tag{3.5.3}$$

Algorithm (3.4.1) requires this constraint to be added to the active set. This implies adding $\underline{n}$ as an extra column to N_m, i.e.

$$N_{m+1} = [N_m \mid \underline{n}], \tag{3.5.4}$$

adding b as an extra element of $\underline{b}_m$, i.e.

$$\underline{b}_{m+1} = \begin{bmatrix} \underline{b}_m \\ \hdashline b \end{bmatrix} , \qquad (3.5.5)$$

forming

$$\underline{\theta}_{m+1}(\underline{x}_u) = \begin{bmatrix} N_m^T \underline{x}_u - \underline{b}_m \\ \hdashline <\underline{n},\underline{x}_u> - b \end{bmatrix} = \begin{bmatrix} \underline{\theta}_m(\underline{x}_u) \\ \hdashline <\underline{n},\underline{x}_u> - b \end{bmatrix} \qquad (3.5.6)$$

computing N_{m+1}^* and

$$\begin{aligned} \underline{x} &= \underline{x}_u - N_{m+1}^{*T} (N_{m+1}^T \underline{x}_u - \underline{b}_{m+1}) \\ &= \underline{x}_u - N_{m+1}^{*T} \underline{\theta}_{m+1}(\underline{x}_u). \end{aligned} \qquad (3.5.7)$$

A similar process is involved when an active constraint, say that corresponding to the last column of N_m and last element of $\underline{b}_m$, is required to be dropped from the active set (see, Step 4, Algorithm (3.4.1)). In this case,

$$N_m = [\, N_{m-1} \mid \underline{n} \,] , \quad \underline{b}_m = \begin{bmatrix} \underline{b}_{m-1} \\ \hdashline b \end{bmatrix} , \quad \underline{\theta}_m(\underline{x}_u) = \begin{bmatrix} \underline{\theta}_{m-1}(\underline{x}_u) \\ \hdashline <\underline{n},\underline{x}_u> - b \end{bmatrix}$$

(3.5.8)

where the constraint to be dropped is

$$<\underline{n}, \underline{x}> - b \geq 0$$

and N_{m-1}, $\underline{b}_{m-1}$, $\underline{\theta}_{m-1}(\underline{x}_u)$, N_{m-1}^* are used to compute the next point

$$\underline{x} = \underline{x}_u - N_{m-1}^{*T} \underline{\theta}_{m-1}(\underline{x}_u) . \qquad (3.5.9)$$

N_m^* may also be used to compute the Lagrange multipliers

$$\begin{aligned} \underline{\lambda}_k &= - (N_m^T H N_m)^{-1} (N_m^T \underline{x}_u - \underline{b}_m) \\ &= - (N_m^T H N_m)^{-1} (N_m^T \underline{x}_u - N_m^T \underline{x}_k) \end{aligned}$$

$$= + (N_m^T H N_m)^{-1} N_m^T H G (\underline{x}_k - \underline{x}_u) \qquad (3.5.10)$$

$$= + N_m^* G (\underline{x}_k - \underline{x}_u) = N_m^* \nabla q(\underline{x}_k) \qquad (3.5.11)$$

An alternative to computing N_m^* and subsequently evaluating $\underline{x}_k$ and $\underline{\lambda}_k$, is using the operator

$$\bar{N}_m = (N_m^T H N_m)^{-1} N_m^T \qquad (3.5.12)$$

with (3.3.6) to compute $\underline{x}_k$ and with (3.5.10) to compute $\underline{\lambda}_k$, i.e.

$$\underline{x}_k = \underline{x}_u - H \bar{N}_m^T \underline{\theta}_m(\underline{x}_u) \qquad (3.5.13)$$

$$\underline{\lambda}_k = (N_m^T H N_m)^{-1} N_m^T H G (\underline{x}_k - \underline{x}_u)$$

$$= \bar{N}_m(\underline{x}_k - \underline{x}_u) . \qquad (3.5.14)$$

Computing $\underline{x}_k$ and $\underline{\lambda}_k$ with either $\bar{N}_m$ or N_m^* as discussed above takes $n^2 + 2mn$ multiplications for both cases.

Updating $\bar{N}_m$ when a column vector, $\underline{n}$, is added to N_m to obtain N_{m+1} is briefly discussed by Fletcher (1969). If N_m, $\underline{n}$ and $\bar{N}_m$ are given, then $\bar{N}_{m+1}$ is required to satisfy the following conditions:

$$\bar{N}_{m+1} H \underline{n}_i = \bar{N}_m H \underline{n}_i = \underline{e}_i \quad , \quad i = 1,2,\dots,m \qquad (3.5.15)$$

where $\underline{n}_i$ and $\underline{e}_i$ are the ith columns of N_m and the $m \times m$ identity matrix respectively,

$$\bar{N}_{m+1} H \underline{n} = \underline{e}_{m+1} \qquad (3.5.16)$$

where $\underline{e}_{m+1}$ is the last column of the $(m+1) \times (m+1)$ identity matrix,

$$\bar{N}_{m+1} H \underline{s} = \underline{o} \quad , \quad \forall \underline{s} \text{ such that } \langle \underline{s}, H \underline{n}_i \rangle = o \ , \ i=1,\dots,m+1. \qquad (3.5.17)$$

The recurrence relation for $\bar{N}_{m+1}$ may be obtained from the above conditions. Since (3.5.17) and also (3.5.15) - (3.5.16) imply that the rank of $\bar{N}_{m+1}$ and $\bar{N}_m$ differ by one, a rank-one modification to $\bar{N}_m$ is required to compute $\bar{N}_{m+1}$. Thus

$$\bar{N}_{m+1} = \begin{bmatrix} \bar{N}_m \\ \hdashline \underline{o}^T \end{bmatrix} + \begin{bmatrix} \underline{v} \\ \hdashline d \end{bmatrix} \underline{u}^T \tag{3.5.18}$$

is the form in which the modification may be expressed. From (3.5.16) it follows that

$$\underline{v} = - \frac{\bar{N}_m H \underline{n}}{\langle \underline{u}, H\underline{n} \rangle} \qquad d = \frac{1}{\langle \underline{u}, H\underline{n} \rangle}$$

where $\underline{v}$ is an m-dimensional and $\underline{u}$ is an n-dimensional vector. From (3.5.15) follows the orthogonality of $H \underline{n}_i$, $i=1,\ldots,m$, to the correction term in (3.5.18). Hence

$$\frac{1}{\langle \underline{u}, H\underline{n} \rangle} \begin{bmatrix} -\bar{N}_m H\underline{n}\underline{u}^T \\ \hdashline \underline{u}^T \end{bmatrix} H \underline{n}_j = \underline{o} \quad , \qquad j=1,\ldots,m.$$

Thus $\underline{u}^T H \underline{n}_j = o$. Consider the operator

$$\bar{P}_m = I - N_m (N_m^T H N_m)^{-1} N_m^T H \tag{3.5.19}$$

that projects every vector $\underline{z} \in E^n$ into a subspace orthogonal to vectors $H N_m \underline{\nu}$ where $\underline{\nu} \in E^m$, i.e.

$$\langle \bar{P}_m \underline{z} , H N_m \underline{\nu} \rangle = o \quad , \quad \forall \underline{z} \in E^n \, , \; \underline{\nu} \in E^m .$$

Alternatively the vector $\bar{P}_m \underline{z}$ is said to be orthogonal to columns of $H N_m$. Hence $\underline{u} = \bar{P}_m \underline{z}$. However, (3.5.17) implies that

$$\langle \bar{P}_m \underline{z}, H \underline{s} \rangle = o$$

which can be assured if $\underline{z} = \underline{n}$. Thus the recurrence formula (3.5.18) may be written as

$$\bar{N}_{m+1} = \begin{bmatrix} \bar{N}_m \\ \text{-----} \\ \underline{o}^T \end{bmatrix} + \begin{bmatrix} -\bar{N}_m H \underline{n} \\ \text{---------} \\ 1 \end{bmatrix} \frac{\underline{u}^T}{<\underline{u}, H\,\underline{n}>} \tag{3.5.20}$$

where $\underline{u} = \bar{P}_m \underline{n}$.

Given $\bar{P}_m$, the recurrence relation for $\bar{P}_{m+1}$ can be obtained using (3.5.20) with (3.5.19). After rearranging terms, this relationship can be expressed as

$$\bar{P}_{m+1} = \bar{P}_m - \frac{\underline{u}\,\underline{u}^T H}{<\underline{u}, H\,\underline{n}>} \; . \tag{3.5.21}$$

When $m=o$, the formulae

$$\bar{N} = \frac{\underline{n}^T}{<\underline{n}, H\underline{n}>} \qquad \bar{P} = I - \frac{\underline{n}\,\underline{n}^T H}{<\underline{n}, H\underline{n}>} \tag{3.5.22}$$

may be used to start the recurrance relations.

Clearly, when the normal $\underline{n}$ is linearly dependent on the columns of N_m, $\underline{n} = N_m \underline{\lambda}$ for an m-vector $\underline{\lambda}$ hence $\bar{P}_m \underline{n} = \underline{o}$. Thus, when $\underline{n}$ is linearly dependent on N_m the denominator, $<\bar{P}_m \underline{n}, H\,\underline{n}>$, in the correction term of (3.5.20) is zero. In this case $\underline{n}$ need not be added to the active set as will be discussed in Section (3.5.3).

When the last column of N_{m+1}, $\underline{n}$, is to be deleted, and given $\bar{N}_{m+1}$, $\bar{N}_m$ is required, (3.5.20) may be used to derive the recurrence relation for $\bar{N}_m$. Thus, rewriting (3.5.20) we have

$$\begin{bmatrix} \bar{N}_m \\ \text{-----} \\ \underline{o}^T \end{bmatrix} = \bar{N}_{m+1} - \begin{bmatrix} -\bar{N}_m H \underline{n} \\ \text{---------} \\ 1 \end{bmatrix} \bar{\underline{n}}^T \tag{3.5.23}$$

where $\bar{\underline{n}} = \underline{u}/<\underline{u}, H\,\underline{n}>$ is the transpose of the m+1st row of $\bar{N}_{m+1}$.

As the equality

$$\bar{N}_m H \underline{\bar{n}} = \bar{N}_m H \bar{P}_m \frac{\underline{n}}{<\bar{P}_m \underline{n}, H \underline{n}>} = \underline{o}$$

follows from the definitions of $\bar{N}_m$, $\bar{P}_m$ and $\underline{\bar{n}}$, multiplying (3.5.23) by $H \underline{\bar{n}}$ yields

$$\begin{bmatrix} - \bar{N}_m H \underline{n} \\ \hline 1 \end{bmatrix} = \frac{\bar{N}_{m+1} H \underline{\bar{n}}}{<\underline{\bar{n}} , H \underline{\bar{n}}>}$$

whence (3.5.23) may be written as

$$\bar{N}_m = \bar{N}_{m+1} - \frac{\bar{N}_{m+1} H \underline{\bar{n}} \underline{\bar{n}}^T}{<\underline{\bar{n}}, H \underline{\bar{n}}>} \tag{3.5.24}$$

The corresponding relation for $\bar{P}_m$ may be derived using (3.5.24) with (3.5.19) to yield

$$\bar{P}_m = \bar{P}_{m+1} + \frac{N_{m+1} \bar{N}_{m+1} H \underline{\bar{n}} \underline{\bar{n}}^T H}{<\underline{\bar{n}}, H \underline{\bar{n}}>}$$

and since $N_{m+1} \bar{N}_{m+1} H \underline{\bar{n}} = \underline{\bar{n}}$, this simplifies to

$$\bar{P}_m = \bar{P}_{m+1} + \frac{\underline{\bar{n}} \underline{\bar{n}}^T H}{<\underline{\bar{n}}, H \underline{\bar{n}}>} . \tag{3.5.25}$$

If $\underline{x}_k$ is the projection of $\underline{x}_u$ onto m+1 hyperplanes and is given by

$$\underline{x}_k = \underline{x}_u - H \bar{N}_{m+1}^T \underline{\theta}_{m+1}(\underline{x}_u) , \tag{3.5.26}$$

and if at $\underline{x}_k$ a constraint was dropped from the active set, then the projection of $\underline{x}_u$ onto the remaining m constraints

$$\underline{x}_{k+1} = \underline{x}_u - H \bar{N}_m \underline{\theta}_m(\underline{x}_u) \qquad (3.5.27)$$

can be expressed in terms of the update in (3.5.24) and the multiplier of the dropped constraint. Eliminating $\underline{x}_u$ between (3.5.26) and (3.5.27)

$$\underline{x}_{k+1} - \underline{x}_k = H(\bar{N}_{m+1}(N_{m+1}^T \underline{x}_u - \underline{b}_{m+1}) - \bar{N}_m(N_m^T \underline{x}_u - \underline{b}_m)), \qquad (3.5.28)$$

since $N_{m+1}^T \underline{x}_k = \underline{b}_{m+1}$, $N_m^T \underline{x}_k = \underline{b}_m$, the recurrance relation (3.5.25) for $\bar{P}_m$ is used to yield

$$\underline{x}_{k+1} - \underline{x}_k = \frac{H \underline{\bar{n}}\, \underline{\bar{n}}^T (\underline{x}_u - \underline{x}_k)}{<\underline{\bar{n}}, H \underline{\bar{n}}>} \qquad (3.5.29)$$

where it was assumed, for simplicity, that the constraint dropped corresponds to the m+1st column of N_{m+1} . As $\underline{\bar{n}}^T$ is the m+1st row of $\bar{N}_m$, by (3.5.14) $\underline{\bar{n}}^T(\underline{x}_k - \underline{x}_u)$ is the m+1st element of the Lagrange multiplier $\underline{\lambda}_k$. Hence

$$\underline{x}_{k+1} - \underline{x}_k = - \frac{H \underline{\bar{n}} \lambda_k^{m+1}}{<\underline{\bar{n}}, H \underline{\bar{n}}>} . \qquad (3.5.30)$$

Thus, $\underline{x}_{k+1} - \underline{x}_k$ lies along $H \underline{\bar{n}}$. A constraint is dropped from the active set when its Lagrange multiplier is negative. With $\lambda_k^{m+1} < o$ and $\underline{\bar{n}} = \bar{P}_m \underline{n} / < \bar{P}_m \underline{n} , H \underline{n}>$ we have

$$<\underline{n}, \underline{x}_{k+1} - \underline{x}_k> = - \frac{<\underline{n}, H \bar{P}_m \underline{n}> \lambda_k^{m+1}}{<\underline{n}, H \bar{P}_m \underline{n}><\underline{n}, H \underline{\bar{n}}>} = - \frac{\lambda_k^{m+1}}{<\underline{\bar{n}}, H \underline{\bar{n}}>} > o . \qquad (3.5.31)$$

Thus, when a constraint with a negative Lagrange multiplier is dropped, the projection of $\underline{x}_u$ onto the remaining hyperplanes in the active set satisfies the dropped constraint as a strict inequality. This also confirms a more direct proof of this result given in Section (3.3.2).

The component of $\nabla q(\underline{x}_k)$ along the direction $\underline{x}_{k+1} - \underline{x}_k$ can be measured, using $H\,\bar{\underline{n}}$, to be

$$< \nabla q(\underline{x}_k) \,,\, H\,\bar{\underline{n}} > \;=\; < \underline{x}_k - \underline{x}_u \,,\, \bar{\underline{n}} > \;=\; \lambda_k^{m+1} \;<\; o\,. \tag{3.5.32}$$

Hence, as in (3.3.51), $\underline{x}_{k+1} - \underline{x}_k$ is a descent direction.

As (3.4.8), in Step 3 of Algorithm (3.4.1), the maximum step size along $H\,\bar{\underline{n}}$, before transgressing a constraint strictly satisfied at $\underline{x}_k$, may be computed using

$$\alpha_i = (b_i - < \underline{n}_i \,,\, \underline{x}_k >)/ < \underline{n}_i \,,\, H\,\bar{\underline{n}} > \,,\quad \forall\, i \in S(\underline{x}_k) \,,\, < \underline{n}_i \,,\, H\,\bar{\underline{n}} > \;<\; o\,. \tag{3.5.33}$$

If the smallest α_i is greater than $-\lambda_k^{m+1}/ < \bar{\underline{n}} \,,\, H\,\bar{\underline{n}} >$, then $\underline{x}_{k+1}$ is a feasible point which is also the minimum in the intersection of the m active constraints. Otherwise, the smallest α_i is set equal to τ in (3.4.9).

The formulae for updating N_m^* can be derived from (3.5.20) and (3.5.24) since

$$N_m^* = \bar{N}_m\, H\,. \tag{3.5.34}$$

The recurrence relations for N_m^* have also been given by Fletcher (1971). When the normal $\underline{n}$ is added to N_m as the m+1st column as in (3.5.4), N_{m+1}^* is given by

$$N_{m+1}^* = \begin{bmatrix} N_m^* \\ \hline \underline{o}^T \end{bmatrix} + \begin{bmatrix} -N_m^*\,\underline{n} \\ \hline 1 \end{bmatrix} \frac{\underline{v}^T}{< \underline{v} \,,\, \underline{n} >} \tag{3.5.35}$$

where

$$\underline{v} = P_m[H]\,\underline{n}$$

$$P_m[H] = H\,\bar{P}_m \tag{3.5.36}$$

and

$$P_{m+1}[H] = P_m[H] - \frac{\underline{v}\,\underline{v}^T}{<\underline{v},\underline{n}>} \quad . \qquad (3.5.37)$$

When the last column of N_{m+1} is removed, to obtain N_m as in (3.5.8), N_m^* and $P_m[H]$ are given by

$$N_m^* = N_{m+1}^* - \frac{N_{m+1}^* G \underline{n}^* \underline{n}^{*T}}{<\underline{n}^*, G\underline{n}^*>} \qquad (3.5.38)$$

$$P_m[H] = P_{m+1}[H] + \frac{\underline{n}^* \underline{n}^{*T}}{<\underline{n}^*, G\underline{n}^*>} \qquad (3.5.39)$$

where $\underline{n}^{*T}$ is the m+1st row of N_{m+1}^* corresponding to the hyperplane dropped from the active set.

The recurrance relationships for $\bar{N}_m$, $\bar{P}_m$ involve H, the inverse of the second derivative matrix of $q(\underline{x})$, whereas the corresponding formulae for N_m^*, $P_m[H]$ involve the second derivative matrix G .

Formulae (3.5.35) and (3.5.37) for computing N_{m+1}^* and $P_{m+1}[H]$ involve $2(n^2 + mn + n)$ operations (counting only multiplications and divisions) without the symmetry of $\underline{v}\,\underline{v}^T$ being taken into account. Formulae (3.5.38) and (3.5.39) for computing N_m^* and $P_m[H]$ involve the same number of operations as above, without the symmetry of $\underline{n}^*\underline{n}^{*T}$ being taken into account. When symmetry is accounted for, one n^2 term in the above count is reduced to $(n^2 + n)/2$. The formulae (3.5.20) and (3.5.22) for computing $\bar{N}_{m+1}$ and $\bar{P}_{m+1}$ involve $(3n^2 + 2mn + 2n)$ operations and the corresponding number for (3.5.24) and (3.5.25), for computing $\bar{N}_m$, $\bar{P}_m$, is $2(n^2 + mn + n)$. In the next section an alternative update is considered.

3.5.2 Updating $(N_m^T H N_m)^{-1}$

The $m \times m$ matrix $(N_m^T H N_m)^{-1}$ occurs in (3.4.5) when computing $\underline{\lambda}_{k+1}$ in Step 2 of Algorithm (3.4.1). Given $(N_m^T H N_m)^{-1}$ and $\underline{\theta}_m(\underline{x}_u)$, the computation of $\underline{x}_{k+1}$ using

$$\underline{\lambda}_{k+1} = -(N_m^T H N_m)^{-1} \underline{\theta}_m(\underline{x}_u) \tag{3.4.5}$$

$$\underline{x}_{k+1} = \underline{x}_u + H N_m \underline{\lambda}_{k+1} \tag{3.4.6}$$

takes $m^2 + mn + n^2$ multiplications. As $m \le n$ this compares favourably with $n^2 + 2mn$ given in Section (3.5.1) for computing the same values using N_m^* or $\bar{N}_m$.

When a constraint normal, $\underline{n}$, is added to the active set $(N_{m+1}^T H N_{m+1})^{-1}$ is given by

$$(N_{m+1}^T H N_{m+1})^{-1} = ([N_m \,\vdots\, \underline{n}]^T H [N_m \,\vdots\, \underline{n}])^{-1} \tag{3.5.40}$$

$$= \left[\begin{array}{c|c} N_m^T H N_m & N_m^T H \underline{n} \\ \hline \underline{n}^T H N_m & \underline{n}^T H \underline{n} \end{array}\right]^{-1}$$

$$= \left[\begin{array}{c|c} A_{11} & A_{12} \\ \hline A_{21} & A_{22} \end{array}\right]^{-1}$$

$$= \left[\begin{array}{c|c} B_{11} & B_{12} \\ \hline B_{21} & B_{22} \end{array}\right]$$

with

$$A_o = A_{22} - A_{21} A_{11}^{-1} A_{12} \tag{3.5.41,a}$$

$$B_{22} = A_o^{-1} \tag{3.5.41,b}$$

$$B_{21} = -A_o^{-1} A_{21} A_{11}^{-1} = -B_{22} A_{21} A_{11}^{-1} \tag{3.5.41,c}$$

$$B_{12} = -A_{11}^{-1} A_{12} A_o^{-1} = -A_{11}^{-1} A_{12} B_{22} \tag{3.5.41,d}$$

$$B_{11} = A_{11}^{-1} + A_{11}^{-1} A_{12} A_o^{-1} A_{21} A_{11}^{-1} = A_{11} + B_{12} A_o B_{21} . \tag{3.5.41,e}$$

Although (3.5.40) is written for only one column being added to N_m, (3.5.41) can be used when more than one column has to be added. In this case $\underline{n}$ is replaced by the matrix whose columns are the normals to be added to N_m as extra columns. When only one column is added (3.5.41) may be written as

$$A_o = \langle \underline{n}, H\,\underline{n} \rangle - \langle \underline{n}, H\,N_m(N_m^T\,H\,N_m)^{-1}\,N_m^T\,H\,\underline{n} \rangle = \langle \underline{n}, P_m[H]\underline{n} \rangle \tag{3.5.42,a}$$

$$B_{22} = \frac{1}{\langle \underline{n}, P_m[H]\underline{n} \rangle} \tag{3.5.42,b}$$

$$B_{21} = -\frac{n^T H\,N_m(N_m^T H\,N_m)^{-1}}{\langle \underline{n}, P_m[H]\underline{n} \rangle} \tag{3.5.42,c}$$

$$B_{12} = -\frac{(N_m^T H\,N_m)^{-1}\,N_m^T\,H\,\underline{n}}{\langle \underline{n}, P_m[H]\,\underline{n} \rangle} \tag{3.5.42,d}$$

$$B = (N_m^T H\,N_m)^{-1} + \frac{(N_m^T H\,N_m)^{-1}\,N_m^T H\,\underline{n}\,\underline{n}^T H\,N_m(N_m^T H\,N_m)^{-1}}{\langle \underline{n}, P_m[H]\,\underline{n} \rangle}\,. \tag{3.5.42,e}$$

In view of (3.5.42,a), when $\underline{n}$ is linearly dependent on N_m, $A_o = o$.

When the constraint corresponding to the m+1 column of N_{m+1} is removed from the active set, $(N_m^T H\,N_m)^{-1}$ may be expressed in terms of $(N_{m+1}^T\,H\,N_{m+1})^{-1}$, using (3.5.41,e)

$$(N_m^T\,H\,N_m)^{-1} = A_{11}^{-1} = B_{11} - B_{12}\,A_o\,B_{21}\,. \tag{3.5.43}$$

The number of multiplications and divisions necessary for computing $(N_{m+1}^T\,H\,N_{m+1})^{-1}$ using (3.5.42) is

$$n^2 + 2m^2 + mn + 2m + n + 1 \tag{3.5.44}$$

if the computation of $\underline{\lambda}_{k+1}$ and $\underline{x}_{k+1}$ are accounted for, $m^2 + mn + n^2$ is added to this number to yield

$$2n^2 + 3m^3 + 2mn + 2m + n + 1 \quad . \tag{3.5.45}$$

When a constraint is dropped, the number of multiplications and divisions required to compute $(N_m^T H N_m)^{-1}$ from (3.5.43) is m+1. As this is smaller than (3.5.44), only the computation involved when adding a constraint to the active set will be considered.

When (3.5.9) and (3.5.11) are used to compute $\underline{x}_{k+1}$ and $\underline{\lambda}_{k+1}$, $n^2 + 2mn$ multiplications are required given N_m^*. When N_{m+1}^* is updated as a constraint normal is added or dropped, another $2n^2 + 2mn + 2m^2$ multiplications and divisions are involved. Hence the total becomes

$$3n^2 + 4nm + 2m^2 \quad . \tag{3.5.46}$$

Since $m \leq n$, (3.5.45) is strictly less than (3.5.46) for all $n \geq o$. Thus, updating $(N_m^T H N_m)^{-1}$ and computing $\underline{x}_{k+1}$, $\underline{\lambda}_{k+1}$ with (3.4.5) and (3.4.6) is strictly better than updating N_m^* and using (3.5.9) and (3.5.11). This conclusion applies even when the symmetry of matrices are accounted for. Furthermore, computing (3.3.32) and (3.3.33) require $2n^2 + mn$ scalar multiplications which is inferior to the amount required by (3.5.9) and (3.5.11) and will not be considered further.

The aim of this section was to establish the comparative superiority of using $(N_m^T H N_m)^{-1}$. However, this has to be weighed against the numerical stability of methods based on (3.3.23) - (3.3.24) or (3.3.13) - (3.3.14) using the orthogonal factorisation of N_m (see

Gill and Murray (1978), Goldfarb (1975)). Among the related operators in these methods are $(Z^T G Z)$ and $(Q_1 H Q_1^T)$. The $L D L^T$ Cholesky factors of these are computed for solving the linear equations discussed in Section (3.3.1). The updates to these Cholesky factors, as constraint normals are added to or dropped from N_m, are extensively discussed in Gill and Murray (1978) and Goldfarb (1975).

3.5.3 Linear Dependence

The constraint

$$\langle \underline{n}, \underline{x} \rangle - b \geq 0 \tag{3.5.47}$$

which is linearly dependent on N_m, i.e. the constraint normals in the active set, becomes active when more than one constraint becomes active at the same time. Let N_m be the matrix of the active constraint normals after the first constraint that has just become active has been added to N_{m-1} as an extra column. The next constraint (and subsequent constraints, if any) that has just been activated may be linearly dependent on N_m, i.e.

$$\underline{n} = N_m \underline{\alpha} \tag{3.5.48}$$

for an m-vector $\underline{\alpha}$ with at least one nonzero element.

The linear dependence of $\underline{n}$ on N_m can be detected while updating the matrix operators discussed in Sections (3.5.1) and (3.5.2) for adding $\underline{n}$ to the active set. It was shown in these sections that while updating N_m^*, the scalar product in the denominator of (3.5.35)

$$\langle \underline{v}, \underline{n} \rangle = \langle \underline{n}, P_m[H]\,\underline{n} \rangle = o \qquad (3.5.49)$$

if $\underline{n}$ is linearly dependent on N_m and hence may be expressed as (3.5.48). Similarly, while computing $\bar{N}_{m+1}$ the denominator of (3.5.20) is given by

$$\langle \underline{u}, H\,\underline{n} \rangle = \langle \bar{P}_m\,\underline{n}, H\,\underline{n} \rangle = \langle \underline{n}, P_m[H]\,\underline{n} \rangle \ . \qquad (3.5.50)$$

Hence the same applies to (3.5.50) when $\underline{n}$ is linearly dependent on N_m. When computing $(N_{m+1}^T\ H\ N_{m+1})^{-1}$, the scalar A_o in (3.5.42,a) is the same inner product as (3.5.49) and the linear dependence of $\underline{n}$ can again be detected. When using one of the methods in Section (3.3.1) based on the orthogonal decomposition of N_m, (1.1.79) - (1.1.80), the relationships $Z^T\ N_m = Q_2\ N_m = o$ hold. Hence, if $n = N\underline{\alpha}$, $Q_2\,\underline{n} = Z^T\underline{n} = \underline{o}$ which may be used to detect the linear dependence of $\underline{n}$ on N_m.

When $\underline{n}$ is linearly dependent on N_m, it need not be added as an extra column of N_m since any movement from the current point, $\underline{x}_k$ to $\underline{x}_{k+1}$ that satisfies

$$N_m^T\,(\underline{x}_{k+1} - \underline{x}_k) = \underline{o}$$

also satisfies

$$\langle \underline{n}, \underline{x}_{k+1} - \underline{x}_k \rangle = o$$

hence (3.5.47) is satisfied as an equality constraint at $\underline{x}_{k+1}$. It may only be necessary to add (3.5.47) to the active set when, after dropping a constraint from the active set, $\underline{n}$ is no longer dependent on the normals of the active constraints.

3.6 CONVERGENCE

Algorithm (3.4.1) for positive definite quadratic programming generates a sequence of points that aim to satisfy the optimality conditions (3.3.2) - (3.3.3) for a global minimum of an equality constrained problem (3.3.1). The equality constraints are those currently in the active set and a member of the above sequence is only prevented from being a global minimum if this minimum violates a previously strictly satisfied constraint (i.e. a constraint that is satisfied as a strict inequality and hence not a member of the current active set). This leads to the following Rule.

Rule (3.6.1)

Constraints are added to the active set only when the current projection of $\underline{x}_u$, $\underline{x}_{k+1}$ (3.4.6), violates or activates it. Furthermore, constraints whose normals are linearly dependent on the existing columns of N_m are not added to the active set (see Section (3.5.3)).

When a constraint is added to the active set, the projection of $\underline{x}_u$ onto the m+1 hyperplanes in the active set lies on a descent direction from the current point. Since from (3.4.6)

$$\underline{x}_{k+1} = \underline{x}_u - H\, N_{m+1}(N_{m+1}^T\, H\, N_{m+1})^{-1}\, (N_{m+1}^T\, \underline{x}_u - \underline{b}_{m+1}) \qquad (3.6.2)$$

with

$$N_{m+1}^T\, \underline{x}_k = \underline{b}_{m+1} \quad \text{and} \quad N_{m+1}^T\, H\, G(\underline{x}_u - \underline{x}_k) = -\, N_{m+1}^T\, H\, \nabla q(\underline{x}_k) \; ,$$

(3.6.2) becomes

$$\underline{x}_{k+1} = \underline{x}_u - H\, N_{m+1}(N_{m+1}^T\, H\, N_{m+1})^{-1}\, N_{m+1}^T\, H\, \nabla q(\underline{x}_k) \; . \qquad (3.6.3)$$

It follows from (3.3.11) that

$$\underline{x}_u = \underline{x}_k - H\,\nabla q(\underline{x}_k)$$

which may be used with (3.6.3) to yield

$$\underline{x}_{k+1} = \underline{x}_k - P_{m+1}[H]\nabla q(\underline{x}_k) \tag{3.6.4}$$

where $P_m[H]$ was defined in (3.5.36). Thus

$$- < \nabla q(\underline{x}_k) \, , \, P_{m+1}[H]\nabla q(\underline{x}_k) > \; < o$$

and $\underline{x}_{k+1}$ lies on the descent direction $-P_{m+1}[H]\nabla q(\underline{x}_k)$ from $\underline{x}_k$, provided $\nabla q(\underline{x}_k) \neq \underline{o}$ or $P_{m+1}\nabla q(\underline{x}_k) \neq \underline{o}$. Note that if $\nabla q(\underline{x}_k) = \underline{o}$, $\underline{x}_k$ is the unconstrained optimum of $q(\underline{x})$ by (3.1.3), and if $P_{m+1}[H]\nabla q(\underline{x}_k) = \underline{o}$, by (3.3.33), there is no feasible descent direction from $\underline{x}_k$ in the intersection of the m+1 constraints. At this stage, the Lagrange multipliers of the m+1 constraints are checked, as in Step 4, Algorithm (3.4.1), and the active constraint with the most negative multiplier is dropped from the active set. This leads to Rule (3.6.5).

<u>Rule</u> (3.6.5)

A constraint is dropped from the active set only when the current point is the global minimum of $q(\underline{x})$ in the intersection of the active constraint hyperplanes and the Lagrange multiplier corresponding to this constraint is negative.

It was shown in (3.5.31) - (3.5.32) that the negativity of the multiplier ensures that the projection of $\underline{x}_u$ onto the remaining active constraints lies on a descent direction from the current point and that this projection also satisfies the constraint that was just dropped. In fact, dropping all constraints with negative multipliers from the active set and projecting $\underline{x}_u$ onto the remaining active constraints generates a descent direction from the current point on which the projection of $\underline{x}_u$ lies. The fact that this is a descent direction can be verified by replacing the subscript m+1 in (3.6.2) - (3.5.4) by m - t where t is the number of constraints with negative multipliers removed from the active set.

Algorithm (3.4.1) seeks solutions, $\underline{x}_c$, with the following properties:

$$G(\underline{x}_c - \underline{x}_u) - N_m \underline{\lambda}_c = \underline{o} \qquad (3.6.6,a)$$

$$\underline{\lambda}_c = -(N_m^T H N_m)^{-1} (N_m^T \underline{x}_u - \underline{b}_m) \geq \underline{o} \qquad (3.6.6,b)$$

$$N_{m_o} \underline{x}_c - \underline{b}_{m_o} \geq \underline{o} . \qquad (3.6.6,c)$$

Theorem (3.3)

The point $\underline{x}_c$ is the global constrained minimum of the positive definite quadratic programming problem (3.1.7) if and only if $\underline{x}_c$ is also the projection of $\underline{x}_u$, the unconstrained optimum of $q(\underline{x})$, onto the feasible region R (3.1.8) .

Proof

The equivalence of (3.6.6) with Kuhn-Tucker conditions has been demonstrated in Sections (3.3.1) - (3.3.2). The fact that $\underline{x}_c$ is the projection of $\underline{x}_u$ has been proved in Theorem (3.1).

The sufficiency of (3.6.6) can be demonstrated by writing for $\underline{x}, \underline{x}_c \in R$, $\underline{x} \neq \underline{x}_c$,

$$<\underline{x} - \underline{x}_u , G(\underline{x} - \underline{x}_u)> - <\underline{x}_c - \underline{x}_u , G(\underline{x}_c - \underline{x}_u)> \geq 2<\underline{x} - \underline{x}_c , G(\underline{x}_c - \underline{x}_u)> \tag{3.6.7}$$

which follows from the positive definiteness of G. Since $\underline{x}_c$ is the projection of $\underline{x}_u$

$$<\underline{x} - \underline{x}_c , G(\underline{x}_c - \underline{x}_u)> \geq o , \tag{3.6.8}$$

by Lemma (4.2) and Luenberger (1969, p. 69). Thus

$$<\underline{x} - \underline{x}_u , G(\underline{x} - \underline{x}_u)> - <\underline{x}_c - \underline{x}_u , G(\underline{x}_c - \underline{x}_u)> \geq o .$$

It should be noted that (3.6.8) could also be arrived at by substituting for $G(\underline{x}_c - \underline{x}_u)$ from (3.6.6,a) and using (3.6.6,b) with, $N_m^T \underline{x}_c = \underline{b}_m$ and $N_m^T \underline{x}_c \geq \underline{b}_m$. Thus, $N_m^T (\underline{x} - \underline{x}_c) \geq \underline{o}$ and

$$<\underline{x} - \underline{x}_c , N_m \underline{\lambda}_c> \geq o$$

which is equivalent to (3.6.8).

The necessity of (3.6.6) may be proved by considering that if (3.6.6,a) is not satisfied, another point may be located in the intersection of the same active constraints which corresponds to a lower value of the objective function. If (3.6.6,b) is not satisfied, a descent direction can be obtained by dropping the inequality constraint with the most negative Lagrange multiplier. Along this direction a point with lower objective function value will be located. Algorithm (3.4.1) follows the above reasoning to locate $\underline{x}_c$. □

Theorem (3.4)

Algorithm (3.4.1) which incorporates Rules (3.6.1) and (3.6.5) converges to the global constrained optimum, $\underline{x}_c$, in a finite number of iterations.

Proof

An iteration means, as in Gill and Murray (1978), the sequence of steps in Algorithm (3.4.1) for which N_m, the matrix of active constraint normals, is fixed (i.e. a change in N_m defines a new iteration).

It was established earlier in this section that Algorithm (3.4.1) generates descent directions by dropping constraints under (3.6.5) and adding constraints under Rule (3.6.1). Thus, the value of the objective function strictly decreases whenever a step is taken. Correspondingly, successive projections of $\underline{x}_u$ generate strictly lower function values. The only exception to this is when a constraint linearly dependent on the current active set has to be added to the active set as an active constraint is dropped. Only in this case a zero step length may be encountered. This happens if a constraint corresponding to a nonzero element of $\underline{\alpha}$ in (3.5.48) is dropped.

The objective function strictly decreases whenever a step with nonzero length is taken. If in this case a constraint is dropped from the current active set, the same intersection of current active constraints will never be returned to since, according to Rule (3.6.5), a constraint is dropped from the active set only if the current point is the minimum of $q(\underline{x})$ in this intersection. As there can be at most n linearly independent constraint hyperplanes in the active set and since there are a finite number of possible combinations of the m_0 inequality constraints, the number of intersections from which the algorithm can drop a constraint from is finite and thus the algorithm terminates in a finite number of steps. □

3.7 CONCLUDING REMARKS

Algorithm (3.4.1) is intended mainly for solving positive definite quadratic programming problems. However, extensions of this algorithm are also discussed. The extensions discussed in Section (3.4.3), for nonlinearly constrained problems, are intended for deterministic optimal control problems with nonlinear models. Such problems are discussed in Chapter 6 where, because of the convenient structure of the model being used, another algorithm is developed.

Algorithm (3.4.1) was implemented on the I C L 1904S at Queen Mary College. The numerical results obtained from this implementation are discussed in the following appendix.

APPENDIX: NUMERICAL RESULTS

Algorithm (3.4.1 was implemented with the updating procedures in Section (3.5.2) on the ICL1904S at Queen Mary College. Its performance was measured against that of the quadratic programming subroutine H02AAF in the Numerical Algorithms Group (NAG) Library. Subroutine H02AAF is the implementation of Beale's algorithm (see, e.g. Beale (1967)).

The test problems considered were the following:

1. minimise $9 - 8x_1 - 6x_2 - 4x_3 + 2x_1^2 + 2x_2^2 + x_3^2 + 2x_1x_2 + 2x_1x_3$

   subject to $x_1 + x_2 + 2x_3 \leq 3$

   $x_1, x_2, x_3 \geq 0$

   Initial Feasible point: $x_1 = 0$, $x_2 = 1$, $x_3 = 1$

   Solution: $x_1 = 1.3333$, $x_2 = .7778$, $x_3 = .4444$

   Source: Beale (1967), Zoutendijk (1970)

2. minimise $9 - 8x_1 - 6x_2 - 4x_3 + 2x_1^2 + 2x_2^2 + x_3^2 + 2x_1x_2$

   subject to the same constraints as Problem 1

   Initial Feasible point: $x_1 = 0$, $x_2 = 1$, $x_3 = 1$

   Solution: $x_1 = 1.4286$, $x_2 = .4286$, $x_3 = .5714$

   Source: Problem 1 with a modified objective function.

3. minimise $- 18x_1 - 16x_2 - 22x_3 - 20x_4 + 3x_1^2$

$$+ x_1x_2 + 8x_1x_3 + 5x_2^2 + x_2x_3$$

$$+ 4x_2x_4 + 8.5x_3^2 + 3x_3x_4 + 5.5x_4^2$$

subject to

$$- x_1 - x_2 - x_3 - x_4 \geq - 1\tfrac{2}{3}$$

$$- 5x_1 \quad - 10x_3 \geq - 2$$

$$- 4x_2 \quad - 5x_4 \geq - 3$$

$$x_1, x_2, x_3, x_4 \geq 0$$

Initial Feasible Point: $x_1 = .4$, $x_2 = .25$, $x_3 = 0$, $x_4 = .4$

Solution: $x_1 = .4$, $x_2 = .2331$, $x_3 = 0$, $x_4 = .4135$

Source: Theil and van de Panne (1960)

4. minimise $2x_1 - 3x_2 + 2x_3 + 20x_1^2 + 5x_2^2 + 5.5x_3^2 + x_2x_3$

subject to

$$- 5x_1 \quad + 2x_3 \geq - 2$$

$$4x_1 + 2x_2 + x_3 \geq 3$$

$$2x_1 + x_2 + x_3 = 2$$

$$x_1, x_2, x_3 \geq 0$$

Initial Feasible Point: $x_1 = .5$, $x_2 = 0$, $x_3 = 1$

Solution: $x_1 = .3127$, $x_2 = .9866$, $x_3 = .3880$

Source: H02 AAF test problem in the NAG library documentation.

5. minimise $42x_1^2 + 9.5x_2^2 - 46x_1 - 25x_2 + 20x_1x_2 + 26$

subject to

$$2x_1 + x_2 \geq 1$$

$$-9x_1 - 2x_2 \geq -6$$

$$x_1, x_2 \geq o$$

Initial Feasible Point: $x_1 = o$, $x_2 = 1$

Solution: $x_1 = .31\ 271$, $x_2 = .98\ 662$

Source: Obtained from problem 4 by eliminating x_3 using the equality constraint.

This problem is an example for the case in which the unconstrained minimum of the objective function is feasible and hence it also is the constrained minimum.

6. minimise $-2x_1 - 2x_2 + 2x_1^2 - 2x_1x_2 + 2x_2^2$

subject to

$$-x_1 - x_2 \geq -1$$

$$x_1 - 6x_2 \geq -2$$

$$x_1, x_2 \geq o$$

Initial Feasible Point: $x_1 = 1$, $x_2 = o$

Solution: $x_1 = .5714$, $x_2 = .4286$

Source: van de Panne and Whinston (1969)

Each quadratic programming problem was solved 30 times and each time the algorithms were reinitialised with the original input. Each such run was repeated at least three times to compute an average value for the total time taken by 30 solutions of each problem. Table 1 below gives this average time taken by either algorithm to solve these problems 30 times

Problem	Number of Variables	Number of Constraints	Average time taken to solve each problem 30 times (milliseconds)	
			Algorithm (3.4.1)	NAG Subroutine H02AAF
1	3	4	262	1310
2	3	4	1048	1048
3	4	7	1310	1573
4	3	6	1048	2097
5	2	4	629	1048
6	2	4	1048	1048

<u>Table 1:</u> Performance of Algorithm (3.4.1) compared with that of NAG quadratic programming algorithm H02AFF

CHAPTER 4

OPTIMISATION WITH LINEAR AND NONLINEAR CONSTRAINTS

4.1 INTRODUCTION

In this chapter a projection algorithm is discussed for solving the nonlinear programming problem

$$\min \{f(\underline{x}) \mid \underline{g}(\underline{x}) \geq \underline{o}\} \qquad (4.1.1)$$

where f and the elements of $\underline{g}$ are differentiable functions given in (1.1.2). The feasible region

$$R \triangleq \{\underline{x} \in E^n \mid \underline{g}(\underline{x}) \geq \underline{o}\} \qquad (4.1.2)$$

will be assumed to be convex or even linear inequality constrained. Furthermore, $f(\underline{x})$ is assumed to be bounded below on R.

In Section (4.1.1) a projection algorithm is discussed. This algorithm can be interpreted as a generalisation of the Goldstein-Levitin-Polyak (GLP) algorithm which computes arcs on R using the steepest descent direction (1.2.22) (Goldstein (1964), Levitin and Polyak (1966)). Algorithm (4.1.7) uses more general projections and Newton or quasi-Newton type descent directions in which second derivatives are approximated with variable metric formulae. The steplength strategies suggested for the GLP algorithm are often difficult to implement. Goldstein (1964), Levitin and Polyak (1966) suggest choosing the steplength from a range bounded by a Lipschitz constant (see Section (4.1.2)). McCormick and Tapia (1972) require univariate minimisation along the steepest descent arc (1.2.23). Bertsekas (1976) suggests an Armijo-type stepsize rule that requires the projection of the unconstrained step for each test value of the stepsize (see Section (4.1.2)). The determination of the stepsize is simplified in algorithm (4.1.7) by dividing this process into two stages. The first determines the length of the unconstrained step. This stepsize can be chosen to be any value in the range (0,2) and determines the position of a possibly infeasible vector $\underline{x}_u$. The second stage is the determination of the stepsize from the range (0,1] that shortens the projected step if the projection of $\underline{x}_u$ on R does not reduce the objective function sufficiently.

In Section (4.1.2) variable metric formulae for updating approximations to the inverse Hessian are discussed. For linear constraints the effect of updating the inverse Hessian approximation and computing the projection of $\underline{x}_u$ is compared with methods that update projection operators involving the inverse Hessian and use these operators to compute descent directions in the intersection of the active constraints.

In Section (4.2.1) alternative stepsize strategies are discussed for existing implementations of (4.1.7). The stepsizes suggested for (4.1.7) are also given in this section.

In Section (4.2.2) the global convergence of (4.1.7) is established when the exact Hessian of $f(\underline{x})$ is available. Superlinear and quadratic rates of convergence of (4.1.7) are also discussed in this section. In Section (4.2.3), algorithm (4.1.7) is shown to converge when variable metric formulae are being used to compute Hessian approximations. Conditions are derived under which the algorithm achieves unit steplengths. The superlinear convergence rate of the algorithm is discussed.

4.1.1 Projection Algorithms for Nonlinear Programming

The basic definition of a quadratic approximation given by (1.1.67) needs to be extended to

$$q_k(\underline{x}_j) \triangleq f(\underline{x}_k) + \langle \nabla f(\underline{x}_k), \underline{x}_j - \underline{x}_k \rangle + \tfrac{1}{2}\langle \underline{x}_j - \underline{x}_k, G_k(\underline{x}_j - \underline{x}_k) \rangle \qquad (4.1.3)$$

where $f(\underline{x})$ is the nonlinear function being approximated by $q_k(\underline{x})$ at $\underline{x}_k$, G_k is the second derivative matrix of $f(\underline{x})$ evaluated at $\underline{x}_k$ (i.e. $[G_k]_{t\ell} = [\partial^2 f/\partial x_t \partial x_\ell]_{\underline{x}=\underline{x}_k}$) and $q_k(\underline{x}_j)$ is the approximation evaluated at $\underline{x}_j$. The matrix G_k may also be approximated by a positive definite matrix $\hat{G}_k$. This will be discussed in Sections (4.1.2) and (4.2.3).

For positive definite G_k, the unconstrained minimum of $q_k(\underline{x})$ is clearly at

$$\underline{x}_u = \underline{x}_k - G_k^{-1} \nabla f(\underline{x}_k) \qquad (4.1.4)$$

satisfying $\nabla q_k(\underline{x}_u) = \underline{o}$. In general $\underline{x}_u$ might violate some constraints in (4.1.2). Thus, if $\underline{x}_u \notin R$ the projection of $\underline{x}_u$ onto R can be computed by solving

$$\min \{ \tfrac{1}{2}\| \underline{x} - \underline{x}_u \|_G^2 \mid \underline{x} \in R \} \qquad (4.1.5)$$

where $G = G_k$ or $G = \hat{G}_k$ with $\hat{G}_k$ denoting a symmetric positive definite approximation to G_k. In the latter case $\hat{G}_k$ also replaces

G_k in (4.1.4). When $G = G_k$ and $\alpha_k = 1$, the vector $\underline{x}$ that solves the projection problem (4.1.5) is the same as

$$\arg\min \{q_k(\underline{x}) \mid \underline{x} \in R\} \tag{4.1.6}$$

since $\nabla q_k(\underline{x}) - \nabla q_k(\underline{x}_u) = \nabla q_k(\underline{x}) = G_k(\underline{x} - \underline{x}_u)$ establishes the correspondence of the Kuhn-Tucker conditions associated with the objective functions of (4.1.5) and (4.1.6). Following Theorem (3.1), the remaining Kuhn-Tucker conditions are identical since both (4.1.5) and (4.1.6) are bounded by R. In algorithm (4.1.7) below the approximations $\hat{G}_k$ and $\hat{H}_k = \hat{G}_k^{-1}$ are used. The special case of (4.1.7) in which the exact second derivatives are available can be derived by replacing $\hat{G}_k$, $\hat{H}_k$ below with G_k, $H_k = G_k^{-1}$ respectively.

<u>Algorithm</u> (4.1.7)

<u>Step 0:</u> Given the starting values $\underline{x}_o$, $\hat{H}_o$ (or $\hat{G}_o$) and the scalars c_1, c_2 with $o < c_2 < c_1 < 1$, $1 \le \bar{\alpha} < 2$;

set $k = o$

<u>Step 1:</u> Compute the unconstrained descent direction

$$\underline{d}_k = - \hat{H}_k \nabla f(\underline{x}_k) \tag{4.1.8}$$

and set the steplength α_k such that $o < \alpha_k \le \bar{\alpha}$. (In unconstrained optimisation (see Powell (1975)) α_k is also required to satisfy

$$< \nabla f(\underline{x}_k + \alpha_k \underline{d}_k), \underline{d}_k > \ \ge \ c_1 < \nabla f(\underline{x}_k), \underline{d}_k > \tag{4.1.9}$$

$$f(\underline{x}_k + \alpha_k \underline{d}_k) \le f(\underline{x}_k) + c_2 \alpha_k < \nabla f(\underline{x}_k), \underline{d}_k > . \tag{4.1.10}$$

Although α_k in this algorithm may also be chosen to satisfy these conditions, it will be shown in Sections (4.2.2)-(4.2.3) that any choice satisfying $\alpha_k \in (0,\bar{\alpha}]$ is sufficient to ensure convergence. A further discussion on possible choices of α_k is given in Section (4.2.1)).

<u>Step 2:</u> Compute

$$\underline{x}_u = \underline{x}_k + \alpha_k \underline{d}_k \; . \tag{4.1.11}$$

If $\underline{x}_u \in R$ set $\underline{x}_p = \underline{x}_u$ and go to Step 4; otherwise go to Step 3 .

<u>Step 3:</u> Compute the projection of $\underline{x}_u$ onto R, denoted by $\underline{x}_p$, i.e.

$$\underline{x}_p = \arg\min \{ \tfrac{1}{2} \| \underline{x} - \underline{x}_u \|^2_{\hat{G}_k} \mid \underline{x} \in R \} \; . \tag{4.1.12}$$

<u>Step 4:</u> Compute a stepsize τ_k, $0 < \tau_k \leq 1$, by one of the methods in Section (4.2.1). This stepsize is for ensuring the sufficient decrease of the objective function in the direction $\underline{x}_p - \underline{x}_k$. Set

$$\underline{x}_{k+1} = \underline{x}_k + \tau_k (\underline{x}_p - \underline{x}_k) \; . \tag{4.1.13}$$

<u>Step 5:</u> If $\underline{x}_{k+1}$ satisfies a prescribed convergence criterion, stop; otherwise go to Step 6.

<u>Step 6:</u> Compute $\hat{H}_{k+1}$ (or $\hat{G}_{k+1}$), set $k = k + 1$ and go to Step 1 .

Clearly the form of (4.1.8) is not a necessary restriction. If $\hat{G}_k$ and not $\hat{H}_k$ is being computed in Step 6 then the equation $\hat{G}_k \underline{d}_k = -\nabla f(\underline{x}_k)$ can be solved for $\underline{d}_k$. Similarly, either $\hat{G}_k$ or $\hat{H}_k$ can be used for solving (4.1.12). When R is linearly constrained (4.1.12) is a quadratic programming problem. Solutions to this problem are discussed in Chapter 3 . Solutions to (4.1.12) when R is non-linearly constrained are discussed in Section (3.4.3). It should be noted that $\underline{x}_p$ in (4.1.12) is unique when R is convex and $\hat{G}_k$ is positive definite. The assumptions on R and $\hat{G}_k$ will be discussed in Sections (4.2.2) - (4.2.3).

Following the equivalence of (4.1.5) and (4.1.6) for $\alpha_k = 1$, (4.1.12) may be interpreted as the minimisation of a quadratic approximation of $f(\underline{x})$. This is summarised in the following theorem.

Theorem (4.1)

For $\alpha_k = 1$, (4.1.12) is equivalent to the solution of the constrained minimisation of a quadratic approximation of $f(\underline{x})$ at $\underline{x}_k$ given by

$$f(\underline{x}_k) + \langle \nabla f(\underline{x}_k) , \underline{x} - \underline{x}_k \rangle + \tfrac{1}{2} \langle \underline{x} - \underline{x}_k , \hat{G}_k(\underline{x} - \underline{x}_k) \rangle . \tag{4.1.14}$$

Proof

When $\alpha_k = 1$, $\underline{x}_u$ in (4.1.11) is the unconstrained minmum of (4.1.14). Thus,

$$\arg\min \{ f(\underline{x}_k) + \langle \nabla f(\underline{x}_k) , \underline{x} - \underline{x}_k \rangle + \tfrac{1}{2} \langle \underline{x} - \underline{x}_k , \hat{G}_k(\underline{x} - \underline{x}_k) \rangle \mid \underline{x} \in R \} \tag{4.1.15}$$

and (4.1.12) are equivalent by Theorem (3.1). The constant term $f(\underline{x}_k)$ in (4.1.14) does not affect the position of the solution vector. □

A special case of Algorithm (4.1.7) may be obtained by setting $\hat{G}_k = I$. This may be seen as an extension of the steepest descent method of unconstrained optimisation to constrained optimisation. In this sense, $\hat{G}_k = I$ may also be considered as a generalisation of Rosen's (1960) method for nonlinear programming with linear inequality constraints.

If R is constrained only by linear inequalities, the algorithm described above does not confine $\underline{x}_k$, $\underline{x}_p$ and $\underline{x}_{k+1} = \underline{x}_k + \tau_k(\underline{x}_p - \underline{x}_k)$, $0 < \tau_k \le 1$, to lie on the same face of the constraint polytope R. Thus, the possibility of $\underline{x}_p - \underline{x}_k$ to cut through the region R is not restricted. Naturally, if the unconstrained step is feasible, its projection is itself and, as stated in Step 2 of (4.1.7), the projection step need not be computed. The projection algorithms for linear inequality constraints (e.g. Goldfarb (1969), Gill and Murray (1972,1973)) discussed in Chapter 1, restrict the successive points, $\underline{x}_k$ and $\underline{x}_{k+1}$, to lie on the same face of R. The algorithm due to Murtagh and Sargent (1969), for linear and nonlinear constraints, takes an unconstrained step. This step is projected onto the local (i.e. at $\underline{x}_k$) linear approximation of the constraints it violates. The method in this section employs a different type of projection subproblem which projects the unconstrained step

$$\underline{x}_u - \underline{x}_k = \alpha_k \underline{d}_k$$

onto the feasible region R. For linear constraints, given $\hat{G}_k$ and the unconstrained step $\alpha_k \underline{d}_k$, consider the projection of $\alpha_k \underline{d}_k$ onto the intersection of the active constraints at $\underline{x}_k$. Treating these constraints as equalities, let the set

$$\Omega$$

denote their intersection. Thus,

$$R \supset \Omega$$

and the inequality

$$\min\{\tfrac{1}{2}\|\underline{x} - \underline{x}_k - \alpha_k \underline{d}_k\|^2_{\hat{G}_k} \mid \underline{x} \in R\} \leq$$

$$\min\{\tfrac{1}{2}\|\underline{x} - \underline{x}_k - \alpha_k \underline{d}_k\|^2_{\hat{G}_k} \mid \underline{x} \in \Omega\} \qquad (4.1.16)$$

holds. The left side denotes subproblem (4.1.12) of algorithm (4.1.7) and the right side denotes the projection steps associated with the other algorithms discussed above. With $\alpha_k \equiv 1$ and

$$\underline{d}_k = - \hat{H}_k \nabla f(\underline{x}_k)$$

expanding the objective function in the above subproblems yields (4.1.14), ignoring the constant terms. Thus, the subproblem on the left has a solution with a lower value for the quadratic approximation (4.1.14) than the subproblem on the right.

During the early iterations of algorithm (4.1.7) the additional step size procedure for τ_k may be necessary. This is designed to force convergence from poor starting points. If the solution of the projection step of the algorithm reduces the objective function sufficiently and satisfies either of the following for $\bar{\alpha} \in [1,2)$

$$f(\underline{x}_p) - f(\underline{x}_k) \leq c_1 \langle \nabla f(\underline{x}_k), \underline{x}_p - \underline{x}_k \rangle, \quad 0 < c_1 < 1 - \frac{\bar{\alpha}}{2}, \qquad (4.1.17)$$

$$f(\underline{x}_p) - f(\underline{x}_k) \leq c_1 (\langle \nabla f(\underline{x}_k), \underline{x}_p - \underline{x}_k \rangle + \tfrac{1}{2} \langle \underline{x}_p - \underline{x}_k, G(\underline{x}_p - \underline{x}_k)\rangle), \quad 0 < c_1 < 1, \qquad (4.1.18)$$

the intermediate step is ignored by setting $\tau_k = 1$ below. The matrix G in (4.1.18) is set to the Hessian or its approximation used in the unconstrained step and the projection subproblem of the algorithm. One of the above conditions is chosen for the algorithm. If this chosen condition fails, $\tau_k \in (0,1]$ is computed by one of the methods discussed in Section (4.2.1) and $\underline{x}_{k+1}$ is set to
$\underline{x}_{k+1} = \underline{x}_k + \tau_k(\underline{x}_p - \underline{x}_k)$.

The strategies for choosing α_k and τ_k are interdependent and will be discussed in Section (4.2.1). Clearly, choosing α_k small brings $\underline{x}_p$ nearer to $\underline{x}_k$ thus making the assumed quadratic approximation to $f(\underline{x})$, between $\underline{x}_k$ and $\underline{x}_p$, more likely to be valid. This validity would ensure that condition (4.2.17) or (4.2.18) is satisfied and $\tau_k = 1$. Hence the computation of τ_k can be avoided. Thus, the chances for the quadratic approximation to hold between $\underline{x}_k$ and $\underline{x}_p$ will be increased by choosing α_k small, provided it is also large enough to ensure adequate progress at each iteration. It will be shown in Sections (4.2.2) - (4.2.3) that provided α_k is chosen to lie within the range (0,2), it does not affect the convergence of the algorithm as long as τ_k is variable. The converse is also true and algorithms have been suggested with $\tau_k = 1$, $\hat{H}_k = I$ in (4.1.7) and various strategies for choosing α_k . Goldstein (1964), Levitin and Polyak (1966) have proposed the case in which α_k is chosen from a range bounded by a Lipschitz constant. This limitation on α_k was later resolved by McCormick and Tapia (1972) who proposed that α_k be determined by univariate minimisation (1.2.23) and by Bertsekas (1976) who suggested the determination of α_k with an Armijo (1966) type step size rule. The constructive approach used in

Sections (4.2.2) and (4.2.3) explains the basis for choosing the strategies for α_k , τ_k and determines the conditions under which α_k , τ_k can be taken to be unity.

4.1.2 Approximations to the Inverse Hessian of the Objective Function and the Computation of Constrained Descent Directions

The problem of generating approximations to the second derivative matrix of the objective function arises from difficulties in using exact second derivatives in optimisation algorithms. These may be outlined as follows:

(i) When the starting point, $\underline{x}_o$, is too far from the optimum, methods based on exact second derivatives called Newton methods may fail to converge even in the unconstrained case. Second order rate of convergence of Newton methods has been proven for functions with a second derivative matrix positive definite and Lipschitz continuous at the optimum, provided $\underline{x}_o$ is sufficiently close to the solution (see, e.g., Ortega and Rheinboldt (1970, Theorem 10.2.2); Dennis and More (1977)).

(ii) When $\underline{x}_o$ is far away from the solution, restrictions (4.1.9), (4.1.10) on the step sizes, α_k , have been suggested (Wolfe (1969), Powell (1975)) to help convergence. Even with such precautions, a well defined step may be computed which is orthogonal to $\nabla f(\underline{x}_k)$. This results in a zero step length hindering further progress. Powell (1966) gives such an example. These difficulties may be overcome either by restricting the steplength to the region in which the quadratic approximation of the objective function is valid (see, e.g., Fletcher (1972)) or by forcing the second derivative matrix to be positive when it is not. When the second derivative approximation is kept positive definite, then unless $\nabla f(\underline{x}_k) = \underline{o}$, $\underline{d}_k = -\hat{H}_k \nabla f(\underline{x}_k)$ is a descent direction.

(iii) Calculating second derivatives is generally not considered desirable since it involves too much work on the computer.

In constrained optimisation, forcing the second derivative matrix approximation, $\hat{G}_k$, to be positive definite may be a serious problem. An objective function with a singular second derivative matrix need not have an unconstrained optimum. Its optimum may exist on some constraints if the Hessian of the Lagrangian (1.2.17) possesses a positive definite projection on the intersection of these constraints. In the unconstrained case, the second derivative matrix is positive definite in the neighbourhood of a strong local minimum. The same need not be true at the constrained optimum and the convergence rate may be affected by forcing $\hat{G}_k$ to be positive definite when it is not. For the linearly constrained case, Gill and Murray (1973) suggest a method restricting only the "projection" of the second derivative matrix onto the set of active constraints, $(Z^T G_k Z)$, to be positive definite (see, (1.1.43) - (1.1.49)). In the linear equality constrained sub-problem (1.1.44), when $(Z^T G_k Z)$ is positive definite and $Z^T \nabla f(\underline{x}_k) = \underline{o}$, $\underline{x}_k$ is a strong local minimum. This implies that the second derivative matrix is restricted to be positive definite in the intersection of the active constraints only. The algorithm in Section (4.1.1) projects the points generated by the unconstrained step onto the inequality constraints rather than taking steps along a descent direction projected onto the active constraints. Thus, for linear inequality constraints this algorithm may generate successive points on different faces of the constraint polytope. This may imply different active sets for successive points. When $\hat{G}_k$ is updated at these points it collects curvature information in all directions. This contrasts with methods proceeding in the intersection of an active set and updating $\hat{G}_k$ at points generated on this set. The curvature information thus collected applies to the directions permitted in the intersection of the active constraints. Gill and Murray (1974,b) show that methods projecting their search direction onto a set of active constraints cannot, in general, collect curvature information in the directions of the normals of these constraints even if they update $\hat{H}_k$ and not $(Z^T G_k Z)^{-1}$ nor

$$P[\hat{H}_k] \triangleq \hat{H}_k - \hat{H}_k N(N^T \hat{H}_k N)^{-1} N^T \hat{H}_k \qquad (4.1.19)$$

where N is the matrix of linearly independent active constraint normals at $\underline{x}_p$. The curvature information gathered using the algorithm in Section (4.1.1) may be evaluated by considering its projection subproblem. Evidently, projections which are solutions of (4.1.12) are least restricted to remain in the intersection of the constraints active at the current feasible point, $\underline{x}_k$. This type of projection has as much freedom as can possibly be provided, to lie on different faces of the constraint polytope R. Hence, the curvature information gleaned from frequent steps through the feasible region R should be richer than those provided by other methods. It is reasonable to assume that projections on regions constrained by linear inequalities might be more costly than projections on linear equalities. However, the simplicity of the special quadratic programming algorithm described in Chapter 3 helps to keep this cost down.

In the following discussion R will be assumed to be constrained by linear inequalities. The inverse Hessian approximation $\hat{H}_k$ will be updated using $\underline{x}_{k+1}$ computed by the algorithms in Section (4.1.1). This approach will be compared with steps taken in the intersection of the active constraints at $\underline{x}_k$, i.e.

$$\underline{x}_{k+1} = \underline{x}_k - \alpha_k \, P[\hat{H}_k] \, \nabla f(\underline{x}_k) \qquad (4.1.20)$$

or

$$\underline{x}_{k+1} = \underline{x}_k - \alpha_k \, Z(Z^T G_k Z)^{-1} Z^T \nabla f(\underline{x}_k) \; . \qquad (4.1.21)$$

Since the latter step is equivalent to (4.1.20), through the equivalence of (1.1.49) and (1.1.54), only (4.1.20) will be considered. When R is nonlinearly constrained, $\underline{x}_p$ in (4.1.12) is conceptually given by (4.1.20) with the matrix N in $P[\hat{H}_k]$ taken as the linearly independent active constraint normals at $\underline{x}_c$. Clearly, this is a characterisation of (4.1.12) rather than a method for solving the corresponding subproblem. However, when R is constrained with linear inequalities (4.1.20) can be explicitly used as a computational technique.

The updating procedure for $\hat{H}_k$ considered in this section is based on the general formulae introduced by Broyden (1967). One representation for this formula is

$$\hat{H}_{k+1} = \hat{H}_k - \frac{\hat{H}_k \underline{\gamma} \, \underline{\gamma}^T \hat{H}_k}{\langle \underline{\gamma}, \hat{H}_k \underline{\gamma} \rangle} + \frac{\underline{\delta} \, \underline{\delta}^T}{\langle \underline{\delta}, \underline{\gamma} \rangle} + \beta_k \, \underline{\omega} \, \underline{\omega}^T \qquad (4.1.22)$$

where

$$\underline{\omega} = \hat{H}_k \underline{\gamma} - \frac{\underline{\delta}}{\langle \underline{\delta}, \underline{\gamma} \rangle} \langle \underline{\gamma}, \hat{H}_k \underline{\gamma} \rangle , \qquad (4.1.23,a)$$

$$\underline{\gamma} = \nabla f(\underline{x}_{k+1}) - \nabla f(\underline{x}_k) , \qquad (4.1.23,b)$$

$$\underline{\delta} = \underline{x}_{k+1} - \underline{x}_k , \qquad (4.1.23,c)$$

and the scalar β_k is a free parameter. Each choice of β_k generates a different updating procedure. The three well known updating formulae correspond to different choices of β_k. Thus,

$$\beta_k = \frac{\langle \underline{\delta}, \underline{\gamma} \rangle}{\langle \underline{\delta}, \underline{\gamma} \rangle \langle \underline{\gamma}, \hat{H}_k \underline{\gamma} \rangle - \langle \underline{\gamma}, \hat{H}_k \underline{\gamma} \rangle^2} \qquad (4.1.24)$$

yields the rank-one formula (see, e.g., Davidon (1959, Appendix), Broyden (1967)), $\beta_k = o$ reduces (4.1.22) to the original Davidon-Fletcher-Powell (DFP) formula due to Davidon (1959), Fletcher and Powell (1963), whereas

$$\beta_k = 1/\langle \underline{\gamma}, \hat{H}_k \underline{\gamma} \rangle \qquad (4.1.25)$$

gives the Broyden-Fletcher-Goldfarb-Shanno (BFGS) formula, discovered independently by Broyden (1969, 1970), Fletcher (1970), Goldfarb (1970) and Shanno (1970). The BFGS formula is generally acknowledged to be the best choice among the members of the family (4.1.22), (see, Powell (1975)).

Another representation of the Broyden (1967) formula (4.1.22) is

$$H_{k+1} = \left(I - \frac{\underline{\delta}\, \underline{\gamma}^T}{\langle \underline{\delta}, \underline{\gamma} \rangle} \right) H_k \left(I - \frac{\underline{\gamma}\, \underline{\delta}^T}{\langle \underline{\delta}, \underline{\gamma} \rangle} \right) + \frac{\underline{\delta}\, \underline{\delta}^T}{\langle \underline{\delta}, \underline{\gamma} \rangle} + \tilde{\beta}_k \underline{\omega}\, \underline{\omega}^T$$

(4.1.26)

$$\tilde{\beta}_k = \beta_k - \frac{1}{\langle \underline{\gamma}, \hat{H}_k \underline{\gamma} \rangle} . \tag{4.1.27}$$

It follows from (4.1.27) that (4.1.26) is identical to (4.1.22). Also, $\tilde{\beta}_k = o$ corresponds to (4.1.25) yielding the BFGS formula. A desirable aspect of both (4.1.22) and (4.1.26) is that they satisfy the equation

$$\hat{H}_{k+1} \underline{\gamma} = \underline{\delta} \tag{4.1.28}$$

which is good because if $f(\underline{x})$ is a quadratic with the inverse of its second derivative matrix given by H, then $H \underline{\gamma} = \underline{\delta}$. For the quadratic case, Huang (1970) showed that for an unconstrained optimisation algorithm taking steps like (4.1.19) with α_k chosen to minimise

$$f(\underline{x}_k - \alpha_k \hat{H}_k \nabla f(\underline{x}_k)) \tag{4.1.29}$$

along the direction $-\hat{H}_k \nabla f(\underline{x}_k)$, given the starting point $\underline{x}_o$, the sequence of points $\underline{x}_k$ $(k=o,1,2,\ldots)$ is independent of the scalar β_k. The important theorem that this result is also true when $f(\underline{x})$ is a general function was proved by Dixon (1972).

A useful result in constrained optimisation is the relationship of the updates $\hat{H}_k$ and the corresponding updates of the operator (4.1.19). Updating $\hat{H}_k$ then computing $P[\hat{H}_k]$ and (4.1.20) also corresponds to solving

$$\underline{x}_p = \arg\min \{ \tfrac{1}{2} \| \underline{x} - \underline{x}_k + \alpha_k \hat{H}_k \nabla f(\underline{x}_k) \|^2_{\hat{G}_k} \mid \underline{x} \in R \}$$

(4.1.30)

and computing $\underline{x}_{k+1} = \underline{x}_k + \tau_k(\underline{x}_p - \underline{x}_k)$, $\quad 0 < \tau_k \le 1$.

Definition (4.1)

The function

$$U_{\beta_k}[\hat{H}_k] \triangleq \hat{H}_{k+1} \tag{4.1.31}$$

is defined for different values of β_k and $\hat{H}_{k+1}$ is given by (4.1.22) or (4.1.26).

Lemma (4.1)

If $\underline{v}$ is any vector such that the matrix $[H + \tau\, \underline{v}\, \underline{v}^T]$ is non-singular, then the projection operator $P[H + \tau\, \underline{v}\, \underline{v}^T]$ may be expressed as

$$P[H + \tau \underline{v}\, \underline{v}^T] = P[H] + \frac{\tau\, P[H]\, G\underline{v}\, \underline{v}^T\, G\, P[H]}{1 + \tau\, \underline{v}^T N(N^T H N)^{-1} N^T\, \underline{v}} \tag{4.1.32}$$

where $G = H^{-1}$.

Proof

The proof follows the same pattern as that of a similar lemma with $H\,\underline{v}$ replacing the vector $\underline{v}$ above (Powell, 1974, Lemma 6.1) .

Using the Sherman-Morrison formula (Householder, 1964) the relationship

$$[N^T[H + \tau\, \underline{v}\, \underline{v}^T]N]^{-1} = [N^T H N]^{-1} - \frac{\tau[N^T H N]^{-1}\, N^T \underline{v}\, \underline{v}^T\, N[N^T H N]^{-1}}{1 + \tau\, \underline{v}^T\, N[N^T H N]^{-1}\, N^T\, \underline{v}} \tag{4.1.33}$$

is obtained. The required result is obtained by constructing $P[H + \tau \underline{v}\, \underline{v}^T]$ according to (1.1.66) and using (4.1.33)

$$P[H + \tau \underline{v}\,\underline{v}^T] = [H + \tau \underline{v}\,\underline{v}^T]$$

$$- [H + \tau\underline{v}\,\underline{v}^T]\, N[(N^T H N)^{-1} - \frac{\tau (N^T H N)^{-1} N^T \underline{v}\,\underline{v}^T N (N^T H N)^{-1}}{1 + \tau\, \underline{v}^T N (N^T H N)^{-1} N^T \underline{v}}] N^T [H + \tau \underline{v}\,\underline{v}^T]$$

$$= H - H N (N^T H N)^{-1} N^T H$$

$$+ \frac{\tau [I - HN(N^T H N)^{-1} N^T]\, \underline{v}\,\underline{v}^T\, [I - HN(N^T H N)^{-1} N^T]^T}{1 + \tau\, \underline{v}^T N (N^T H N)^{-1} N^T \underline{v}} . \quad \square$$

(4.1.34)

In establishing the relationship of the updates of $\hat{H}_k$ and $P[H_k]$, the special case of $\beta_k = o$ in (4.1.22) will be first considered. The more general cases will be subsequently discussed.

Theorem (4.2)

If the active set at $\underline{x}_{k+1}$, $I(\underline{x}_{k+1})$, is the same as that at $\underline{x}_k$, i.e.

$$I(\underline{x}_{k+1}) = I(\underline{x}_k) \qquad (4.1.35)$$

and if $\underline{\delta}_k$ in (4.1.22) is a multiple of

$$- P[\hat{H}_k]\, \nabla f(\underline{x}_k) \qquad (4.1.36)$$

then

$$P[U_o[\hat{H}_k]] = U_o[P[\hat{H}_k]] \qquad (4.1.37)$$

where the subscript o indicates $\beta_k = o$ in (4.1.22) .

Proof

Setting $\beta_k = o$ in (4.1.22) yields

$$U_o[\hat{H}_k] = \hat{H}_k - \frac{\hat{H}_k \underline{\gamma}\,\underline{\gamma}^T \hat{H}_k}{<\underline{\gamma},\ \hat{H}\,\underline{\gamma}>} + \frac{\underline{\delta}\,\underline{\delta}^T}{<\underline{\delta},\underline{\gamma}>} \quad . \qquad (4.1.38)$$

As $\underline{\delta}$ is a multiple of (4.1.36), the equality

$$N^T \underline{\delta} = \underline{o} \qquad (4.1.39)$$

is satisfied. Using (4.1.38) and (4.1.39), $P[U_o[\hat{H}_k]]$ may be written as

$$P[U_o[\hat{H}_k]] = U_o[\hat{H}_k]$$

$$[\hat{H}_k - \frac{\hat{H}_k \underline{\gamma}\,\underline{\gamma}^T \hat{H}_k}{<\underline{\gamma},\hat{H}_k \underline{\gamma}>}] N [N^T[\hat{H}_k - \frac{\hat{H}_k \underline{\gamma}\,\underline{\gamma}^T H_k}{<\underline{\gamma},\hat{H}_k \underline{\gamma}>}]N]^{-1} N^T[\hat{H}_k - \frac{\hat{H}_k \underline{\gamma}\,\underline{\gamma}^T \hat{H}_k}{<\underline{\gamma},\hat{H}_k \underline{\gamma}>}]$$

$$= \frac{\underline{\delta}\,\underline{\delta}^T}{<\underline{\delta},\underline{\gamma}>} + P[\hat{H}_k - \frac{\hat{H}_k \underline{\gamma}\,\underline{\gamma}\,\hat{H}_k}{<\underline{\gamma},\ \hat{H}_k\underline{\gamma}>}] \quad . \qquad (4.1.40)$$

Using Lemma (4.1) with $\underline{v} = \hat{H}_k \underline{\gamma}$, $H = H_k$ and $\tau = -1/<\underline{\gamma}, \hat{H}_k \underline{\gamma}>$, (4.1.40) is reduced to

$$P[U_o[\hat{H}_k]] = \frac{\underline{\delta}\,\underline{\delta}^T}{<\underline{\delta},\underline{\gamma}>}$$

$$+ P[\hat{H}_k] - \frac{P[\hat{H}_k]\,\underline{\gamma}\,\underline{\gamma}^T\,P[\hat{H}_k]}{<\underline{\gamma},P[\hat{H}_k]\,\underline{\gamma}>} \qquad (4.1.41)$$

$$= U_o[P[\hat{H}_k]]$$

which is the required result. □

Powell (1974) has proved a similar result for the formula (4.1.26), i.e.

$$P[U_o[\hat{H}_k]] = U_o[P[\hat{H}_k]] \qquad (4.1.42)$$

where the subscript o implies $\tilde{\beta}_k = o$ in (4.1.26). Furthermore,

Powell (1974) has proved a more general theorem establishing the relationship

$$P[U_{\tilde{\beta}_k}[\hat{H}_k]] = U_{\beta'_k}[P[H_k]] \qquad (4.1.43)$$

using (4.1.26) where

$$\beta'_k = \frac{\tilde{\beta}_k}{1 + \tilde{\beta}_k \underline{\gamma}\, \hat{H}_k\, N(N^T \hat{H}_k N)^{-1} N^T \hat{H}_k\, \underline{\gamma}} \;. \qquad (4.1.44)$$

This establishes the equivalence of the sequence of matrices generated by updating directly $P[\hat{H}_k]$ and the sequence obtained by updating $\hat{H}_k$ and thereby constructing $P[\hat{H}_k]$. A typical algorithm using the former strategy is due to Goldfarb (1969) and a typical algorithm using the latter is due to Murtagh and Sargent (1969). If the correspondence of $\tilde{\beta}_k$ and β'_k (4.1.44) is satisfied, then provided the same starting point, $\underline{x}_0$, and step lengths α_k $(k=0,1,2,\ldots)$ are used, Powell (1974) shows that algorithms based on either strategy generate the same sequence of points x_k $(k=1,2,\ldots)$.

The equivalence of using the DFP formula for updating $\hat{H}_k$ and constructing $P[\hat{H}_k]$ with Goldfarb's algorithm was established by Theorem (4.2). The same result may be established by setting $\beta_k = 0$ in (4.1.27) and thus

$$\tilde{\beta}_k = -\frac{1}{\langle \underline{\gamma}, \hat{H}_k \underline{\gamma} \rangle} \;. \qquad (4.1.45)$$

Substituting this value in (4.1.44) and using (4.1.19) yields

$$\beta'_k = -\frac{1}{\langle \underline{\gamma}, P[\hat{H}_k]\, \underline{\gamma} \rangle} \;. \qquad (4.1.46)$$

The equivalence of the BFGS formula for both strategies is implied by (4.1.42) or by setting $\tilde{\beta}_k = 0$ in (4.1.44). A similar correspondence may be established for the rank-one formula. Using (4.1.24) with (4.1.26)

$$\tilde{\beta}_k = \beta_k - \frac{1}{\langle \underline{\gamma}, \hat{H}_k \underline{\gamma} \rangle} = \frac{\langle \underline{\delta}, \underline{\gamma} \rangle}{\langle \underline{\delta}, \underline{\gamma} \rangle \langle \underline{\gamma}, \hat{H}_k \underline{\gamma} \rangle - \langle \underline{\gamma}, \hat{H}_k \underline{\gamma} \rangle} - \frac{1}{\langle \underline{\gamma}, H_k \underline{\gamma} \rangle}$$

$$= \frac{1}{\langle \underline{\delta}, \underline{\gamma} \rangle - \langle \underline{\gamma}, \hat{H}_k \underline{\gamma} \rangle} . \tag{4.1.47}$$

Substituting (4.1.47) into (4.1.44) and using (4.1.19) yields

$$\beta_k' = \frac{1}{\langle \underline{\delta}, \underline{\gamma} \rangle - \langle \underline{\gamma}, P[\hat{H}_k] \underline{\gamma} \rangle} , \tag{4.1.48}$$

which is the required result. □

Another property is the independence of the minimising sequence $\underline{x}_k$, $(k = 0,1,2,\ldots)$ from the choice of $\tilde{\beta}_k$ or β_k', even if their relationship (4.1.44) is ignored, provided α_k $(k = 0,1,\ldots)$ is chosen to minimise (4.1.29) subject to the constraints. This follows from (4.1.43) and the original result for the unconstrained case, mentioned above, due to Dixon (1972).

When an inequality constraint is dropped from the active set, methods based on the strategy $U_\beta[P[\hat{H}_k]]$ (e.g. Goldfarb's (1969) algorithm) use the formula

$$P_-[\hat{H}_k] = P[\hat{H}_k] + \frac{P_-[I]\, \underline{n}\, \underline{n}^T P_-[I]}{\langle \underline{n}, P_-[I]\, \underline{n} \rangle} \tag{4.1.49}$$

to update $P[H_k]$. In (4.1.49), $\underline{n}$ is the normal of the constraint being dropped from the active set, $P_-[\hat{H}_k]$ is (4.1.19) with the vector

$\underline{n}$ removed from the columns of N and $P\_[I]$ is $P\_[\hat{H}_k]$ with $\hat{H}_k$ replaced by the identity matrix. Note that the matrix norm

$$\|P\_[\hat{H}_k] - P[\hat{H}_k]\| = \left\| \frac{P\_[I]\,\underline{n}\,\underline{n}^T\,P\_[I]}{<\underline{n},\, P\_[I]\,\underline{n}>} \right\|$$

$$= \text{trace}\left[\frac{P\_[I]\,\underline{n}\,\underline{n}^T\,P\_[I]\,P\_[I]\,n\,n^T\,P\_[I]}{<\underline{n},\, P\_[I]\,\underline{n}>^2}\right]^{\frac{1}{2}}$$

$$= 1 \qquad (4.1.50)$$

since $P^T\_[I] = P\_[I]$ and $P\_[I]P\_[I] = P\_[I]$. Thus, the difference $P\_[\hat{H}_k] - P[\hat{H}_k]$ is negligible when the elements of $P[\hat{H}_k]$ are relatively large. The updates are not scaled naturally. This difficulty does not occur with the alternative strategy, $P[U_\beta[\hat{H}_k]]$, as the matrix $\hat{H}_k$ is not altered to take account of the changes in the active set. For the latter strategy, the updating formulae corresponding to (4.1.49) are discussed in Chapter 3 as the recursion relationship for the operator $P_m[\hat{H}]$.

To account for the addition of a constraint to the active set, the former strategy, $U_\beta[P[\hat{H}_k]]$, updates $P[\hat{H}_k]$ as follows:

$$P_+[\hat{H}_k] = P[\hat{H}_k] - \frac{P[\hat{H}_k]\,\underline{n}\,\underline{n}^T\,P[\hat{H}_k]}{<\underline{n},\, P[\hat{H}_k]\,\underline{n}>}, \qquad (4.1.51)$$

where $\underline{n}$ is the normal of the constraint being added and $P_+[\hat{H}_k]$ is the resulting operator with $\underline{n}$ added to the columns of N. The matrix $P_+[\hat{H}_k]$ has the undesirable property that its rank is one lower than $P[\hat{H}_k]$ and this may result in the loss of useful curvature information. Methods directly updating $\hat{H}_k$ (i.e. the latter strategy, $P[U_\beta[\hat{H}_k]]$) avoid this loss which becomes appearent when $\underline{n}$, in turn, is dropped. As long as $\underline{x}_{k+1}, \underline{x}_{k+2}, \ldots$ remain in a neighbourhood of

$\underline{x}_k$, the curvature information accumulated in $\hat{H}_k$ will remain valid even if $\underline{x}_{k+1}, x_{k+2}, \ldots$ remain on the same active set. If $\underline{x}_{k+1}, \underline{x}_{k+2}, \ldots,$ move away from this neighbourhood then the curvature information is only collected from movements on the intersection of the active constraints. If this intersection does not change the more general curvature information in $\hat{H}_k$ looses its significance in regions outside a neighbourhood of $\underline{x}_k$. Only if constraints are deleted from the active set that curvature information can be obtained in the direction of the normals of the deleted constraints. Thus, algorithms progressing on one face of the constraint polytope can, in general, gather curvature information on this face (see, also, Section (1.1.2)). Full curvature information can be maintained in $\hat{H}_k$ by changing this face frequently and cutting through the feasible region.

4.2 CONVERGENCE OF THE ALGORITHM

The convergence of the Algorithm in Section (4.1.1) depends on the stepsize strategies chosen for α_k and τ_k. These strategies are discussed in Section (4.2.1). The convergence proofs are given in Sections (4.2.2) and (4.2.3).

4.2.1 Stepsize Strategies

The stepsizes α_k in the unconstrained step (4.1.11) and τ_k in

$$\underline{x}_{k+1} = \underline{x}_k + \tau_k(\underline{x}_p - \underline{x}_k) \qquad (4.2.1)$$

determine the convergence of algorithm (4.1.7). The choice of α_k can be made within a range (0,2) provided τ_k is chosen so that $f(\underline{x}_{k+1})$ is sufficiently lower than $f(\underline{x}_k)$. Thus, $f(\underline{x}_{k+1})$ and $f(\underline{x}_k)$ are expected to satisfy

$$f(\underline{x}_k) - f(\underline{x}_{k+1}) \geq \sigma(t_k(\underline{x}_k , \underline{x}_p)) \qquad (4.2.2)$$

where σ is a mapping $\sigma:[0,\infty) \to [0,\infty)$ and $t_k(\underline{x}_k , \underline{x}_p)$ is a function of $\underline{x}_k , \underline{x}_p$ such that the sequence $\{t_k(\underline{x}_k , \underline{x}_p)\} \subset [0,\infty)$. Furthermore, $\lim_{k\to\infty} \sigma(t_k(\underline{x}_k,\underline{x}_p)) = 0$ implies $\lim_{k\to\infty} t_k(\underline{x}_k,\underline{x}_p) = 0$. For algorithm (4.1.7) the two choices for $t_k(\underline{x}_k,\underline{x}_p)$

$$t_k(\underline{x}_k,\underline{x}_p) = -< \nabla f(\underline{x}_k) , \underline{x}_p - \underline{x}_k >$$

$$t_k(\underline{x}_k,\underline{x}_p) = -< \nabla f(\underline{x}_k) , \underline{x}_p - \underline{x}_k > - \tfrac{1}{2} < \underline{x}_p - \underline{x}_k , G(\underline{x}_p - \underline{x}_k) >$$

will be derived in Sections (4.2.2) and (4.2.3) for G denoting the Hessian or its approximation used in the unconstrained and projection steps of the algorithm. Clearly once $t_k(\underline{x}_k,\underline{x}_p)$ is chosen for an algorithm, it is assumed that this is not changed to its alternative during the course of computations.

A general framework is established first to discuss the properties of existing implementations of (4.1.7) with $\hat{H}_k = I$ and alternative convergent stepsize strategies. Finally the strategies used in Sections (4.2.2) and (4.2.3) to establish superlinear and quadratic rates of convergence will be discussed.

<u>Definition (4.2)</u>

The gradient $\nabla f(\underline{x})$ is said to satisfy a Lipschitz condition with constant $M > 0$ on the set $R \subseteq E^n$ if the inequality

$$\| \nabla f(\underline{x}_i) - \nabla f(\underline{x}_j) \| \leq M \| \underline{x}_i - \underline{x}_j \| \qquad (4.2.3)$$

is satisfied for $\underline{x}_i, \underline{x}_j \in R$.

When $\hat{H}_k = I$, $\forall k$, algorithm (4.1.7) reduces to some variant of the steepest descent algorithm. For such simple algorithms a general formula may be derived to show that the sequence $f(\underline{x}_k), f(\underline{x}_{k+1}), \ldots$ is monotonically decreasing. Using the first order Taylor series expansion formula (see, e.g. Polak, 1971, p. 293) the expression

$$\begin{aligned} f(\underline{x}_{k+1}) - f(\underline{x}_k) &= \int_0^1 \langle \nabla f(\underline{x}_k + t(\underline{x}_{k+1} - \underline{x}_k)), \underline{x}_{k+1} - \underline{x}_k \rangle dt \\ &= \langle \nabla f(\underline{x}_k), \underline{x}_{k+1} - \underline{x}_k \rangle \\ &+ \int_0^1 \langle \nabla f(\underline{x}_k + t(\underline{x}_{k+1} - \underline{x}_k)) - \nabla f(\underline{x}_k), \underline{x}_{k+1} - \underline{x}_k \rangle dt \end{aligned} \qquad (4.2.4)$$

is obtained. This expression will be used with the next two lemmas to discuss the properties of three simple implementations of (4.1.7) with simple step size strategies for α_k and τ_k.

<u>Lemma (4.2)</u>

For $\underline{x}_u \in E^n$, its projection, $\underline{x}_p$, on R i.e.

$$\underline{x}_p = \arg\min \{ \|\underline{x} - \underline{x}_u\|_G^2 \mid \underline{x} \in R \}$$

for any symmetric positive definite $n \times n$ matrix G and any vector $\underline{y} \in R$, the relationship

$$\langle \underline{x}_p - \underline{y} , G(\underline{x}_p - \underline{x}_u) \rangle \leq 0 \qquad (4.2.5)$$

holds.

Proof

For $G = I$, (4.2.5) is a well known result (see, e.g. Luenberger (1969), p. 69). We prove that the more general case also holds.

Suppose, contrary to (4.2.5), that $\exists\ \underline{y}_1 \in R$ such that $< \underline{x}_p - \underline{y}_1 , G(\underline{x}_p - \underline{x}_u) > \ = \varepsilon > 0$. Consider $\underline{\omega}(\alpha) = (1-\alpha)\underline{x}_p + \alpha\underline{y}_1$, $0 \leq \alpha \leq 1$. Since R is convex, each $\underline{\omega}(\alpha) \in R$. Also

$$\|\underline{\omega}(\alpha) - \underline{x}_u\|_G^2 = \|(1-\alpha)(\underline{x}_p - \underline{x}_u) + \alpha(\underline{y}_1 - \underline{x}_u)\|_G^2$$

$$= (1-\alpha)^2 \|\underline{x}_p - \underline{x}_u\|_G^2 + 2\alpha(1-\alpha) < \underline{y}_1 - \underline{x}_u , G(\underline{x}_p - \underline{x}_u) >$$

$$+ \alpha^2 \|\underline{y}_1 - \underline{x}_u\|_G^2 .$$

The norm $\|\underline{\omega}(\alpha) - \underline{x}_u\|_G^2$ is a differentiable function of α and its derivative at $\alpha = 0$ is

$$\frac{d}{d\alpha}\|\underline{\omega}(\alpha) - \underline{x}_u\|_G^2 = -2\|\underline{x}_p - \underline{x}_u\|_G^2 + 2 < \underline{y}_1 - \underline{x}_u , G(\underline{x}_p - \underline{x}_u) >$$

$$= -2 < \underline{x}_p - \underline{y}_1 , G(\underline{x}_p - \underline{x}_u) > \ = -2\varepsilon < 0 .$$

Thus, for some small positive α, $\|\underline{\omega}(\alpha) - \underline{x}_u\|_G^2 < \|\underline{x}_p - \underline{x}_u\|_G^2$ which is a contradiction since $\underline{x}_p$ is, by hypothesis, the projection of $\underline{x}_u$ and thereby minimises this norm. Hence, no such $\underline{y}_1$ can exist.

Conversely, suppose that $\underline{x}_p \in R$ is such that (4.2.5) holds $\forall \underline{y} \in R$. Then for any $\underline{y} \in R$, $\underline{y} \neq \underline{x}_p$ we have

$$\|\underline{y} - \underline{x}_u\|_G^2 = \|\underline{y} - \underline{x}_p + \underline{x}_p - \underline{x}_u\|_G^2$$

$$= \|\underline{y} - \underline{x}_p\|_G^2 + 2 < \underline{y} - \underline{x}_p , G(\underline{x}_p - \underline{x}_u) >$$

$$+ \|\underline{x}_p - \underline{x}_u\|_G^2$$

$$> \|\underline{x}_p - \underline{x}_u\|_G^2 .$$

Hence $\underline{x}_p$ is a unique minimising vector. □

Lemma (4.3)

For algorithms with $\underline{x}_u = \underline{x}_k - \alpha_k G^{-1} \nabla f(\underline{x}_k)$ for some positive definite matrix G , the relationship

$$\|\underline{x}_p - \underline{x}_k\|_G^2 \leq -\alpha_k < \nabla f(\underline{x}_k), \underline{x}_p - \underline{x}_k > \qquad (4.2.6)$$

holds with $\underline{x}_p$ denoting the projection of $\underline{x}_u$ defined in Lemma (4.2).

Proof

We have

$$\|\underline{x}_p - \underline{x}_k\|_G^2 = < \underline{x}_p + \alpha_k G^{-1} \nabla f(\underline{x}_k) - \underline{x}_k , G(\underline{x}_p - \underline{x}_k) >$$

$$- \alpha_k < \nabla f(\underline{x}_k), \underline{x}_p - \underline{x}_k > .$$

The first term on the right hand side may be formulated as (4.2.5) whence the result follows. □

Applying (4.2.1) to (4.2.4), the first term of the right hand side may be expressed as

$$< \nabla f(\underline{x}_k), \underline{x}_{k+1} - \underline{x}_k > = \tau_k < \nabla f(\underline{x}_k), \underline{x}_p - \underline{x}_k > . \qquad (4.2.7)$$

Using the Lipschitz condition (4.2.3) and (4.2.6) with (4.2.7), the relationship (4.2.4) is reduced to

$$f(\underline{x}_{k+1}) - f(\underline{x}_k) \leq -\frac{\tau_k}{\alpha_k} \| \underline{x}_p - \underline{x}_k \|^2 + \frac{M\tau_k^2}{2} \| \underline{x}_p - \underline{x}_k \|^2$$

$$= \left(-\frac{\tau_k}{\alpha_k} + \frac{M \tau_k^2}{2} \right) \|\underline{x}_p - \underline{x}_k\|^2 . \qquad (4.2.8)$$

For algorithms such as Goldstein (1964) using the steepest descent direction (i.e. $\hat{H}_k = I$ in (4.1.7)) Levitin and Polyak (1966, Section 5), suggest the step sizes

$$0 < \alpha_k \le \frac{2}{M + 2\varepsilon_1} \;, \quad \varepsilon_1 > 0 \quad \text{with} \quad \tau_k = 1 \;, \forall_k . \qquad (4.2.9)$$

Substituting (4.2.9) in (4.2.8) yields

$$f(\underline{x}_{k+1}) - f(\underline{x}_k) = f(\underline{x}_p) - f(\underline{x}_k) \le (-\frac{1}{\alpha_k} + \frac{M}{2}) \|\underline{x}_p - \underline{x}_k\|^2$$

$$= -\varepsilon_1 \|\underline{x}_p - \underline{x}_k\|^2 . \qquad (4.2.9)$$

which establishes the monotonic decrease of the sequence $\{f(\underline{x}_k)\}$. □

Levitin and Polyak also prove the linear rate of convergence of this algorithm. Another example for such algorithms is Levitin and Polyak's generalisation of Dem'yanov and Rubinov's (1967) method. This is also based on the choice $\hat{H}_k = I$ in (4.1.7) and is known as a conditional gradient method. Levitin and Polyak (1966, Section 6) prove the convergence of the algorithm provided $\alpha_k = 1$ and τ_k is given by (4.2.10). This choice of α_k can easily be extended to the bounds

$$0 < \alpha_k \le \bar{\alpha} < \infty \quad \text{with} \quad \tau_k = \min\{1, \frac{\gamma_k \|<\nabla f(\underline{x}), \underline{x}_k - \underline{x}_p>\|}{\|\underline{x}_k - \underline{x}_p\|^2}\} \qquad (4.2.10)$$

where $0 < \varepsilon_2 \le \gamma_k \le (2 - \varepsilon_3)/M$, $\varepsilon_3 > 0$. Using the general expression (4.2.8) for the family of algorithms with $\hat{H}_k = I$, $f(\underline{x}_k)$ may be shwon to be monotonically decreasing. For $\tau_k = 1$ in (4.2.10) ,

$$f(\underline{x}_{k+1}) - f(\underline{x}_k) \le (\frac{-2 + M\bar{\alpha}}{2\bar{\alpha}}) \|\underline{x}_p - \underline{x}_k\|^2 , \qquad (4.2.11)$$

since $\tau_k = 1$

$$\frac{2 - \varepsilon_3}{M} \ge \gamma_k > \frac{\|\underline{x}_k - \underline{x}_p\|^2}{<\nabla f(\underline{x}_k), \underline{x}_k - \underline{x}_p>} \qquad (4.2.12)$$

and using (4.2.6)

$$-\varepsilon_3 > M\bar{\alpha} - 2$$

which may be substituted in (4.2.11) to yield

$$f(\underline{x}_{k+1}) - f(\underline{x}_k) \;<\; -\; \frac{\varepsilon_3}{2\bar{\alpha}} \;\| \underline{x}_p - \underline{x}_k \|^2 . \qquad (4.2.13)$$

For $\tau_k = \dfrac{\gamma_k < \nabla f(\underline{x}_k), \underline{x}_k - \underline{x}_p >}{\| \underline{x}_k - \underline{x}_p \|^2}$,

$$f(\underline{x}_{k+1}) - f(\underline{x}_k) \;\le\; -\frac{\gamma_k}{\bar{\alpha}} < \nabla f(\underline{x}_k), \underline{x}_k - \underline{x}_p > + \;\frac{M\gamma_k^2}{2} \;\frac{< \nabla f(\underline{x}_k), \underline{x}_k - \underline{x}_p >^2}{\| \underline{x}_k - \underline{x}_p \|^2}$$

Using (4.2.6)

$$f(\underline{x}_{k+1}) - f(\underline{x}_k) \;\le\; -\,\gamma_k \left(\frac{2 - M\gamma_k}{2\,\bar{\alpha}^2} \right) \|\underline{x}_p - \underline{x}_k\|^2$$

and applying the bounds on γ_k along with the first inequality of (4.2.12) the complement of (4.2.13) is obtained

$$f(\underline{x}_{k+1}) - f(\underline{x}_k) \;\le\; -\,\frac{\varepsilon_2 \varepsilon_3}{2\bar{\alpha}^2} \;\| \underline{x}_p - \underline{x}_k \|^2 \qquad (4.2.14)$$

which establishes the monotonic decrease of $\{f(\underline{x}_k)\}$. □

Demyanov and Rubinov's original algorithm employs a different strategy for determining τ_k . This strategy is based on choosing $\alpha_k = 1$ and τ_k as given below. However, α_k can also be chosen from the range

$$0 < \alpha_k \le \bar{\alpha} < \infty \quad \text{with} \quad \tau_k = \arg\min \{ f(\underline{x}_k - \xi(\underline{x}_k - \underline{x}_p)) \mid 0 \le \xi \le 1 \}.$$

This choice is clearly easy to implement and may also be shown to yield monotonically decreasing $f(\underline{x}_k)$. Using (4.2.8) leads to

$$f(\underline{x}_{k+1}) - f(\underline{x}_k) \le f(\underline{x}_k - \xi(\underline{x}_k - \underline{x}_p)) - f(\underline{x}_k)$$

$$\le \;\frac{\xi}{\bar{\alpha}} \left(-1 + \xi \frac{M\bar{\alpha}}{2} \right) \| \underline{x}_p - \underline{x}_k \|^2 . \qquad (4.2.15)$$

Since for $0 \le \xi \le 1$ the inequality (4.2.15) holds, taking $0 \le \xi' \le \min\{1, 2/M\bar{\alpha}\}$ ensures that the expression in brackets is not positive for $\xi = \xi'$. This establishes the monotonic decrease of $\{f(\underline{x}_k)\}$. The linear rate of convergence of Dem'yanov and Rubinov's algorithm is established in Levitin and Polyak (1966, Section 6). □

Bertsekas (1976) suggests the use of algorithm (4.1.7) with $\hat{H}_k = I$ and $\tau_k = 1$ such that $\underline{x}_{k+1} = \underline{x}_p$. The steplength α_k is chosen such that $\alpha_k \geq 0$, $\alpha_k = \beta^{m_k} s$ where $0 < \beta < 1$, $s > 0$ and m_k is the smallest nonnegative integer satisfying

$$f(\underline{x}_{k+1}) - f(\underline{x}_k) \leq c_1 < \nabla f(\underline{x}_k), \underline{x}_{k+1} - \underline{x}_k > , \ 0 < c_1 < \tfrac{1}{2} .$$

Thus, the projection subproblem (4.1.12) has to be computed for all test values of α_k until this condition is satisfied. The stepsize strategies (4.2.21) and (4.2.26) below avoid the computation of new projection subproblems for each value of the step size.

The following result establishes a lower bound on the step size α_k if it is intended to be set at a univariate minimum along the direction $- H \nabla f(\underline{x}_k)$.

Lemma (4.4)

Let the gradient of the objective function $f(\underline{x})$ be Lipschitz continuous on E^n (see Definition (4.2)). If the step size α^*

$$\underline{x}_{k+1} - \underline{x}_k = - \alpha^* H \nabla f(\underline{x}_k) \qquad (4.2.16)$$

is set such that

$$f(\underline{x}_k - \alpha^* H \nabla f(\underline{x}_k)) = \min_{\alpha} f(\underline{x}_k - \alpha H \nabla f(\underline{x}_k)) \qquad (4.2.17)$$

for positive definite H , then

$$\alpha^* \geq \frac{<\nabla \underline{f}(\underline{x}_k), H\, \nabla f(\underline{x}_k)>}{M\, <H\, \nabla f(\underline{x}_k),\, H\, \nabla f(\underline{x}_k)>} \quad . \tag{4.2.18}$$

Proof

Since (4.2.17) implies

$$<\nabla f(\underline{x}_{k+1})\, ,\, \underline{x}_{k+1} - \underline{x}_k> \; = \; o \; ,$$

we can write

$$-<\nabla f(\underline{x}_k)\, ,\underline{x}_{k+1} - \underline{x}_k> \; = \; <\nabla f(\underline{x}_{k+1}) - \nabla f(\underline{x}_k)\, ,\, \underline{x}_{k+1} - \underline{x}_k> \; ,$$

and using (4.2.3) along with the Schwartz inequality

$$\begin{aligned} <\nabla f(\underline{x}_{k+1}) - \nabla f(\underline{x}_k)\, ,\underline{x}_{k+1} - \underline{x}_k> \; &\leq \; \| \nabla f(\underline{x}_{k+1}) - \nabla f(\underline{x}_k) \| \; \| \underline{x}_{k+1} - \underline{x}_k \| \\ &\leq \; M \, \| \underline{x}_{k+1} - \underline{x}_k \|^2 \\ &= \; M <\underline{x}_{k+1} - \underline{x}_k\, , \underline{x}_{k+1} - \underline{x}_k> \\ &= \; M <-\alpha^* \, H \nabla f(\underline{x}_k), \, \underline{x}_{k+1} - \underline{x}_k> \end{aligned}$$

hence $-<\nabla f(\underline{x}_k)\, ,\underline{x}_{k+1} - \underline{x}_k> \; \leq \; -\alpha^* \, M <H\, \nabla f(\underline{x}_k)\, ,\, \underline{x}_{k+1} - \underline{x}_k>$ from which the result follows. □

In the case of the steepest descent direction, $H = I$ and (4.2.18) reduces to

$$\alpha^* \; \geq \; \frac{1}{M} \quad . \tag{4.2.19}$$

Since the Lipschitz constant is not an easily quantifiable value, this bound cannot be used in practice.

When $\hat{H}_k$ in (4.1.7) is the exact inverse Hessian or an approximation to it, the step size strategies for α_k and τ_k are as follows:

(i) (a) Choose $\bar{\alpha} \in [1,2)$ and $\alpha_k \in (o,\bar{\alpha}]$.

The step size restrictions (4.1.9) - (4.1.10) ensure an adequate decrease in the objective function along the unconstrained direction $\underline{d}$ in (4.1.11). As τ_k in Step 4 of both algorithms ensures that $f(\underline{x}_{k+1})$ is lower than $f(\underline{x}_k)$, (4.1.9) - (4.1.10) need not be considered any further. The same argument also applies against adopting the suggestions of Goldstein (1964), Levitin and Polyak (1966), Bertsekas (1976) discussed in Sections (1.2.1) and (4.1.1).

(i) (b) Choose $\bar{\alpha} \in [1,2)$

$$\alpha_k = \underset{o<\alpha<\bar{\alpha}}{\arg\min} \{f(\underline{x}_k - \alpha \hat{H}_k \nabla f(\underline{x}_k))\} \tag{4.2.20}$$

McCormick and Tapia (1972) have also suggested univariate minimisation (1.2.23). For the same reasons given above (4.2.20) also need not be considered.

(ii)(a) Choose $\tau_k \in (o,1]$ as the largest value satisfying the stepsize restriction

$$f(\underline{x}_{k+1}) - f(\underline{x}_k) \leq c_1 < \nabla f(\underline{x}_k), \underline{x}_{k+1} - \underline{x}_k > \tag{4.2.21}$$

where

$$\underline{x}_{k+1} - \underline{x}_k = \tau_k (\underline{x}_p - \underline{x}_k) , \tag{4.2.22}$$

$$o < c_1 < 1 - \frac{\bar{\alpha}}{2} \tag{4.2.23}$$

and $\bar{\alpha} \in [1,2)$ is the above bound α_k.

The inequality (4.2.21) is similar to (4.1.10) and that if $\alpha_k = 1$, $\forall k$, (4.2.23) becomes $o < c_1 < \frac{1}{2}$. Substituting (4.2.22) in (4.2.21) yields

$$f(\underline{x}_{k+1}) - f(\underline{x}_k) \leq \tau_k c_1 < \nabla f(\underline{x}_k), \underline{x}_p - \underline{x}_k > .$$

As a generalisation of the linear approximation bound in the right hand side of (4.2.21) consider the quadratic approximation

$$\bar{q}_k(\underline{x}) \triangleq q_k(\underline{x}) - f(\underline{x}_k) = \langle \nabla f(\underline{x}_k), \underline{x} - \underline{x}_k \rangle + \tfrac{1}{2} \langle x - \underline{x}_k, G_k(\underline{x} - \underline{x}_k) \rangle \tag{4.2.24}$$

where $q_k(\underline{x})$ is defined by (4.1.3) .

Lemma (4.5)

The inequality

$$\bar{q}_k(\underline{x}_{k+1}) \leq \tau_k \, \bar{q}_k(\underline{x}_p) \tag{4.2.25}$$

holds for $\underline{x}_{k+1}$ given by (4.2.22), $o \leq \tau_k \leq 1$ and positive semidefinite G_k .

Proof

$$\begin{aligned} \bar{q}_k(\underline{x}_{k+1}) &= \langle \nabla f(\underline{x}_k), \underline{x}_{k+1} - \underline{x}_k \rangle + \tfrac{1}{2} \langle \underline{x}_{k+1} - \underline{x}_k, G_k(\underline{x}_{k+1} - \underline{x}_k) \rangle \\ &= \tau_k (\langle \nabla f(\underline{x}_k), \underline{x}_p - \underline{x}_k \rangle + \frac{\tau_k}{2} \langle \underline{x}_p - \underline{x}_k, G_k(\underline{x}_p - \underline{x}_k) \rangle) \\ &\leq \tau_k \, \bar{q}_k \, (\underline{x}_p) \, . \end{aligned}$$

□

The result (4.2.25) may be used in designing the quadratic restriction to τ_k .

(ii)(b) Choose $o < \tau_k \leq 1$ as the largest τ_k satisfying the step size restriction

$$f(\underline{x}_{k+1}) - f(\underline{x}_k) \leq c_1 \, \tau_k \bar{q}_k(\underline{x}_p) \tag{4.2.26}$$

where

$$o < c_1 < 1 \, . \tag{4.2.27}$$

In (4.2.24), G_k may also be replaced by $\hat{G}_k$ or $\hat{G}_k^L$. According to each choice of G a new q and hence a new stepsize similar to (ii)(b) will be defined in Section (4.2.3).

(ii)(c) Choose

$$\tau_k = \underset{0 < \tau \leq 1}{\arg\min} f(\underline{x}_k + \tau(\underline{x}_p - \underline{x}_k)) . \tag{4.2.28}$$

4.2.2 Convergence Proofs: Exact Second Derivatives

In this section global convergence results of algorithm (4.1.7) will be discussed assuming that the exact Hessian of $f(\underline{x})$ is available and $\hat{G}_k = G_k$. The scalar $\bar{\alpha}$ is assumed to be chosen such that $\bar{\alpha} \in [1,2)$.

Lemma (4.6)

For $0 < \alpha_k \leq 2$,

$$\bar{q}_k(\underline{x}_p) \leq (\tfrac{1}{2} - \frac{1}{\alpha_k}) \|\underline{x}_p - \underline{x}_k\|^2_{G_k} \leq 0 . \tag{4.2.29}$$

Proof

From (4.2.24) $\bar{q}_k(\underline{x}_p)$ may be written as

$$\bar{q}_k(\underline{x}_p) = \langle \nabla f(\underline{x}_k), \underline{x}_p - \underline{x}_k \rangle + \tfrac{1}{2} \langle \underline{x}_p - \underline{x}_k, G_k(\underline{x}_p - \underline{x}_k) \rangle .$$

Using Lemma (4.3), this may be written as

$$\bar{q}_k(\underline{x}_p) \le -\frac{1}{\alpha_k} \|\underline{x}_p - \underline{x}_k\|_G^2 + \tfrac{1}{2}\|\underline{x}_p - \underline{x}_k\|_{G_k}^2 . \qquad (4.2.30)$$

The result clearly follows for the given range of α_k . □

Corollary

For $o < \alpha_k \le 2$

$$\ell\|\underline{y}\|^2 \le \langle\underline{y}, G_k\,\underline{y}\rangle , \quad \ell > o , \quad \forall_k \text{ and } \forall_{\underline{y}} \in R . \qquad (4.2.31)$$

the bound

$$\ell\,\xi\;\|\underline{x}_p - \underline{x}_k\|^2 \le -\bar{q}_k(\underline{x}_p) , \qquad \xi \ge o \qquad (4.2.32)$$

holds. Furthermore $o < \alpha_k < 2$ implies $\xi > o$.

Proof

Using (4.2.29) the following relationship may be written

$$\bar{q}_k(\underline{x}_p) \le \ell\left(\frac{1}{2} - \frac{1}{\alpha_k}\right)\|\underline{x}_p - \underline{x}_k\|^2 = -\ell\xi\,\|\underline{x}_p - \underline{x}_k\|^2 \qquad (4.2.33)$$

for $\xi = -\left(\frac{1}{2} - \frac{1}{\alpha_k}\right)$. Since it was already established that

$\bar{q}_k(\underline{x}_p) \le o$, the inequality (4.2.32) follows from (4.2.33) with the open range $o < \alpha_k < 2$ implying $\xi > o$. □

Theorem (4.3)

Let $R \subseteq E^n$ be a convex set and $f(\underline{x})$ be a twice continuously differentiable function on R with

$$\ell\|\underline{y}\|^2 \le \|\underline{y}\|_{G_k}^2 = \langle\underline{y}, G_k\,\underline{y}\rangle \le L\|\underline{y}\|^2 , \quad \ell > o , \quad \underline{x}_k , \; \underline{y} \in R, \qquad (4.2.34)$$

$$\underline{x}_u = \underline{x}_k - \alpha_k H_k \nabla f(\underline{x}_k) , \quad o < \alpha_k < 2 , \qquad (4.2.35)$$

and $\underline{x}_p$ denoting the projection of $\underline{x}_u$ onto R, given by (4.1.12), while

$$\underline{x}_{k+1} = \underline{x}_k + \tau_k(\underline{x}_p - \underline{x}_k) , \quad o < \tau_k \leq 1 . \tag{4.2.36}$$

For all the strategies outlined in the previous section to compute τ_k, i.e.,

$$f(\underline{x}_{k+1}) - f(\underline{x}_k) \leq c_1 \, \tau_k \, \bar{q}(\underline{x}_p) , \qquad o < c_1 < 1 , \tag{4.2.37}$$

$$f(\underline{x}_{k+1}) - f(\underline{x}_k) \leq c_1 \, \tau_k < \nabla f(\underline{x}_k) , \underline{x}_p - \underline{x}_k > , \quad o < c_1 < 1 - \frac{\bar{\alpha}}{2} , \tag{4.2.38}$$

$$\tau_k = \underset{o < \tau \leq 1}{\arg\min} \; f(\underline{x}_k + \tau(\underline{x}_p - \underline{x}_k)) , \tag{4.2.28}$$

the corresponding sequences of $f(\underline{x}_k)$, $k = 0,1,\ldots$ are monotonically decreasing.

Proof

Consider the second order Taylor expansion (Polak (1971)),

$$f(\underline{x}_{k+1}) - f(\underline{x}_k) = < \nabla f(\underline{x}_k), \underline{x}_{k+1} - \underline{x}_k > \tag{4.2.39}$$
$$+ \int_0^1 (1 - t) < \underline{x}_{k+1} - \underline{x}_k , G[\underline{x}_k - t(\underline{x}_k - \underline{x}_{k+1}]) (\underline{x}_{k+1} - \underline{x}_k) > dt$$

where $G[.]$ denotes the second derivative matrix of $f(\underline{x})$ evaluated at $[.]$. Furthermore, using (4.2.36),(4.2.25) and the Schwartz inequality, the relationship

$$f(\underline{x}_{k+1}) - f(\underline{x}_k) = < \nabla f(\underline{x}_k) , \underline{x}_{k+1} - \underline{x}_k > + \tfrac{1}{2} < \underline{x}_{k+1} - \underline{x}_k , G_k(\underline{x}_{k+1} - \underline{x}_k) >$$
$$+ \int_0^1 (1 - t) < \underline{x}_{k+1} - \underline{x}_k , \{G[\underline{x}_k - t(\underline{x}_k - \underline{x}_{k+1})] - G_k\}(\underline{x}_{k+1} - \underline{x}_k) > dt$$
$$\leq \bar{q}_k(\underline{x}_{k+1}) + \tau_k^2 \; \mu_k \| \underline{x}_p - \underline{x}_k \|^2 \tag{4.2.40}$$

$$\leq \tau_k \bar{q}_k(\underline{x}_p) + \tau_k^2 \mu_k \| \underline{x}_p - \underline{x}_k \|^2 \tag{4.2.41}$$

is obtained with

$$\mu_k = \int_0^1 (1-t) \| G[\underline{x}_k + \tau_k t(\underline{x}_p - \underline{x}_k)] - G_k \| \, dt \, . \tag{4.2.42}$$

using the bound (4.2.32) with (4.2.41) yields

$$f(\underline{x}_{k+1}) - f(\underline{x}_k) \leq \tau_k \bar{q}_k(\underline{x}_p) \left(1 - \frac{\tau_k \mu_k}{\ell \xi}\right) . \tag{4.2.43}$$

Thus,

$$c_1 \leq 1 - \frac{\tau_k \mu_k}{\ell \xi} \tag{4.2.44}$$

and for given μ_k, ℓ and ξ there exist $o < \tau_k \leq 1$ such that $o < c_1 \leq 1 - \frac{\tau_k \mu_k}{\ell \xi} < 1$. Hence, given c_1 , there exist τ_k that that satisfy (4.2.37). Since $\bar{q}_k(\underline{x}_p) \leq o$, by Lemma (4.6), it follows that $f(\underline{x}_k)$ is monotonically decreasing for (4.2.37) .

To prove that the sequence is monotonically decreasing with (4.2.38), first an upper bound to μ_k (4.2.42) is established using (4.2.34)

$$\mu_k = \int_0^1 (1-t) \| G[\underline{x}_k + \tau_k t(\underline{x}_p - \underline{x}_k)] - G_k \| \, dt \leq L \, . \tag{4.2.45}$$

Writing (4.2.41) with this bound, the inequality

$$\begin{aligned} f(\underline{x}_{k+1}) - f(\underline{x}_k) &\leq \tau_k \bar{q}_k(\underline{x}_p) + \tau_k^2 L \| \underline{x}_p - \underline{x}_k \|^2 \\ &= \tau_k \langle \nabla f(\underline{x}_k), \underline{x}_p - \underline{x}_k \rangle \left(1 + \tau_k \frac{\langle \underline{x}_p - \underline{x}_k, G_k(\underline{x}_p - \underline{x}_k) \rangle}{2 \langle \nabla f(\underline{x}_k), \underline{x}_p - \underline{x}_k \rangle}\right. \\ &\quad \left. + \frac{\tau_k L \| \underline{x}_p - \underline{x}_k \|^2}{\langle \nabla f(\underline{x}_k), \underline{x}_p - \underline{x}_k \rangle}\right) \end{aligned} \tag{4.2.46}$$

is obtained. The inequality (4.2.6) of Lemma (4.3) written as

$$\frac{\ell \, ||\underline{x}_p - \underline{x}_k||^2}{-<\nabla f(\underline{x}_k), \underline{x}_p - \underline{x}_k>} \leq \frac{||\underline{x}_p - \underline{x}_k||^2_{G_k}}{-<\nabla f(\underline{x}_k), \underline{x}_p - \underline{x}_k>} \leq \alpha_k \tag{4.2.47}$$

may be used to simplify (4.2.46) and obtain

$$f(\underline{x}_{k+1}) - f(\underline{x}_k) \leq \tau_k <\nabla f(\underline{x}_k), \underline{x}_p - \underline{x}_k> (1 - \frac{\tau_k \alpha_k}{2} - \frac{\mu_k \tau_k \alpha_k}{\ell}) \tag{4.2.48}$$

$$\leq \tau_k <\nabla f(\underline{x}_k), \underline{x}_p - \underline{x}_k> (1 - \frac{\tau_k \alpha_k}{2} - \frac{L \tau_k \alpha_k}{\ell}) \tag{4.2.49}$$

It is clear from Lemma (4.3) that $<\nabla f(\underline{x}_k), \underline{x}_p - \underline{x}_k> \leq o$. Thus, for $o < \alpha_k < 2$, $o < \ell \leq L$, there exist $o < \tau_k \leq 1$ such that

$$o < c_1 < 1 - \frac{\bar{\alpha}}{2} \quad \text{with}$$

$$o < c_1 \leq 1 - \frac{\tau_k \alpha_k}{2} - \frac{\mu_k \tau_k \alpha_k}{\ell} < 1 - \frac{\bar{\alpha}}{2} \tag{4.2.50}$$

The sequence is hence monotonically decreasing for (4.2.38). The restriction $\alpha_k < 2$ is imposed, instead of $\alpha_k \leq 2$, because of the upper bound on c_1.

Finally, consider the choice of τ_k according to (4.2.28). Let τ_k^* denote the value determined using (4.2.28) and

$$\underline{x}^*_{k+1} = \underline{x}_k + \tau^*_k(\underline{x}_p - \underline{x}_k) \; . \tag{4.2.51}$$

Furthermore, let $\underline{x}_{k+1}$ denote another vector computed using τ_k satisfying condition (4.2.37). Hence

$$f(\underline{x}^*_{k+1}) - f(\underline{x}_k) \leq f(\underline{x}_{k+1}) - f(\underline{x}_k) \leq c_1 \, \tau_k \, \bar{q}_k(\underline{x}_p) \tag{4.2.52}$$

holds for τ_k satisfying (4.2.37). The first inequality in (4.2.52) implies that τ_k^* also satisfies (4.2.37). Thus the sequence generated by computing τ_k with (4.2.28) is also monotonically decreasing. (This last part can also be proved using condition (4.2.38) instead of (4.2.28) with an identical argument). Hence the proof is completed. □

For τ_k satisfying (4.2.37), Lemma (4.3) may be used to express (4.2.43) with (4.2.44) as

$$f(\underline{x}_{k+1}) - f(\underline{x}_k) \leq c_1 \, \tau_k \left(- \frac{1}{\alpha_k} + \frac{1}{2} \right) \| \underline{x}_p - \underline{x}_k \|^2_{G_k} . \qquad (4.2.53)$$

The monotonic decrease of the sequence $f(\underline{x}_k)$, $k = 0,1,2$ is thus ensured for $o < \alpha_k < 2$. The same consideration is applied to condition (4.2.38) for which

$$f(\underline{x}_{k+1}) - f(\underline{x}_k) \leq - \frac{c_1 \tau_k}{\alpha_k} \| \underline{x}_p - \underline{x}_k \|^2_{G_k} \qquad (4.2.54)$$

is obtained. Clearly, the sequence $f(\underline{x}_k)$ decreases monotonically for $o < \alpha_k < 2$. In the next theorem , the convergence of the sequences $\{\bar{q}_k(\underline{x}_p)\}$ and $\{<\nabla f(\underline{x}_k), \underline{x}_p - \underline{x}_k>\}$ is established.

<u>Theorem (4.4)</u>

If $f(\underline{x})$ satisfies

$$\ell \| \underline{y} \|^2 \leq \| \underline{y} \|^2_{G_k} = < \underline{y} , G_k \, \underline{y} > \leq L \| \underline{y} \|^2 , \ \ell > o , \ \underline{y} , \underline{x}_k \in R \qquad (4.2.34)$$

then for $o < \tau_k \leq 1$,

$$\lim_{k \to \infty} \bar{q}_k(\underline{x}_p) = o \qquad (4.2.55)$$

and

$$\lim_{k \to \infty} <\nabla f(\underline{x}_k), \underline{x}_p - \underline{x}_k> = o \qquad (4.2.56)$$

Proof

The first inequality of (4.2.34) implies that $f(\underline{x})$ is bounded from below on R (see, e.g., Polak, 1971, Theorem B.2.8). Thus, using the monotonicity results of Theorem (4.3)

$$\lim_{k\to\infty} f(\underline{x}_{k+1}) - f(\underline{x}_k) = o \tag{4.2.57}$$

is obtained. Since both $\bar{q}_k(\underline{x}_p) \leq o$ and $<\nabla f(\underline{x}_k), \underline{x}_p - \underline{x}_k> \leq o$ by Lemma (4.6) and Lemma (4.3) respectively, (4.2.55) follows from (4.2.43) and (4.2.56) follows from (4.2.48). □

The option $\tau_k = o$ has been excluded from Theorem (4.4) since for the convex region R with $\underline{x}_p \neq \underline{x}_k , \underline{x}_p , \underline{x}_k \in R$, the line segment $[\underline{x}_k , \underline{x}_p] \neq \emptyset$ and $\underline{x}_p - \underline{x}_k$ being a descent direction at $\underline{x}_k$ implies $\tau_k > o$.

Lemma (4.7)

If

$$\lim_{k\to\infty} \bar{q}_k(\underline{x}_p) = o \tag{4.2.55}$$

or if

$$\lim_{k\to\infty} <\nabla f(\underline{x}_k) , \underline{x}_p - \underline{x}_k> = o \tag{4.2.56}$$

then,

$$\lim_{k\to\infty} \|\underline{x}_p - \underline{x}_k\| = o . \tag{4.2.58}$$

Proof

For (4.2.55), inequality (4.2.29) of Lemma (4.6) provides the result. For (4.2.56), inequality (4.2.6) of Lemma (4.3) implies (4.2.58). □

Theorem (4.5)

If $f(\underline{x})$ satisfies (4.2.34), then for algorithms with descent steps (4.2.35), (4.2.36) and restrictions on τ_k (4.2.37), (4.2.38), the relationship

$$f(\underline{x}_c) = \inf \{f(\underline{x}) \mid \underline{x} \in R\} = \lim_{k\to\infty} f(\underline{x}_k) \tag{4.2.59}$$

holds.

Proof

The restriction (4.2.34) implies the convexity of $f(\underline{x})$ (see, Polak, 1971, Theorem B.2.8). Thus, a unique minimum, $\underline{x}_c$, of $f(\underline{x})$ on R exists. Hence,

$$o \le f(\underline{x}_k) - f(\underline{x}_c) \le \langle \nabla f(\underline{x}_k), \underline{x}_k - \underline{x}_c \rangle \tag{4.2.60}$$

$$\begin{aligned}
&= \frac{1}{\alpha_k} \langle \alpha_k H_k \nabla f(\underline{x}_k), G_k(\underline{x}_k - \underline{x}_p + \underline{x}_p - \underline{x}_c) \rangle \\
&\quad + \frac{1}{\alpha_k} \langle \underline{x}_k - \underline{x}_p, G_k(\underline{x}_c - \underline{x}_p) \rangle \\
&\quad - \frac{1}{\alpha_k} \langle \underline{x}_k - \underline{x}_p, G_k(\underline{x}_c - \underline{x}_p) \rangle \\
&= \langle \nabla f(\underline{x}_k), \underline{x}_k - \underline{x}_p \rangle \\
&\quad + \frac{1}{\alpha_k} \langle \underline{x}_k - \underline{x}_p - \alpha_k H_k \nabla f(\underline{x}_k), G_k(\underline{x}_c - \underline{x}_p) \rangle \\
&\quad - \frac{1}{\alpha_k} \langle \underline{x}_k - \underline{x}_p, G_k(\underline{x}_c - \underline{x}_p) \rangle \\
&\le \langle \nabla f(\underline{x}_k), \underline{x}_k - \underline{x}_p \rangle + \frac{1}{\alpha_k} \langle \underline{x}_p - \underline{x}_k, G_k(\underline{x}_c - \underline{x}_p) \rangle
\end{aligned} \tag{4.2.61}$$

where the first inequality of (4.2.60) is due to $\underline{x}_c$, the second inequality is due to the convexity of $f(\underline{x})$ and (4.2.61) is due to

Lemma (4.2). Since $||G_k(\underline{x}_c - \underline{x}_p)||$ is bounded for convex R and f, the result follows from Theorem (4.4) and Lemma (4.7). □

Next we consider rates of convergence.

Theorem(4.6)

For $f(\underline{x})$ satisfying (4.2.34) and the steps (4.2.35),(4.2.36) with τ_k chosen according to (4.2.37) or (4.2.38) or (4.2.28), $\{\underline{x}_k\}$ converges Q-superlinearly to the unique constrained minimum $\underline{x}_c$ on the convex set R. If, furthermore, $||G[\underline{x}] - G[\underline{y}]|| \leq M_G||\underline{x} - \underline{y}||$ holds for $M_G > o$ and $\underline{x}, \underline{y} \in R$, then $\{\underline{x}_k\}$ is quadratically convergent to $\underline{x}_c$.

Proof

The proof is of three parts, depending on the criterion used for selecting τ_k. First

$$f(\underline{x}_{k+1}) - f(\underline{x}_k) \leq c_1 \tau_k \bar{q}_k(\underline{x}_p), \quad o < c_1 < 1, \tag{4.2.37}$$

will be considered. It follows from the result, (4.2.58), of Lemma (4.7) that the scalar μ_k given by equation (4.2.42) tends to zero, i.e.

$$\lim_{k\to\infty} \mu_k = \lim_{k\to\infty} \int_o^1 (1 - t) ||G[\underline{x}_k + t\tau_k(\underline{x}_p - \underline{x}_k)] - G_k|| dt = o. \tag{4.2.62}$$

Hence there exists an integer $K_\alpha(c_1)$ such that for any $o < c_1 < 1$ and $k \geq K_\alpha(c_1)$ the bound (4.2.44) becomes

$$o < c_1 \leq 1 - \frac{\tau_k \mu_k}{\ell\xi} \leq 1 - \frac{\mu_k}{\ell\xi} < 1, \quad o < \tau_k \leq 1.$$

This implies that (4.2.37) is satisfied for $\tau_k = 1$, $k \geq K_\alpha(c_1)$. Similarly, since $\xi = -(\frac{1}{2} - \frac{1}{\alpha_k})$ there exists an integer $K(c_1)$ for which the bound (4.2.44) becomes

$$o < c_1 \leq 1 - \frac{2\,\tau_k \mu_k}{\ell} \leq 1 - \frac{2\,\mu_k}{\ell} < 1\,; k \geq K(c_1), \alpha_k = 1, o < \tau_k \leq 1.$$

This implies that (4.2.37) is satisfied with $\tau_k = \alpha_k = 1$ for $k \geq K(c_1)$.

Consider the quadratic approximation

$$\bar{q}_k(\underline{x}_{k+1}) = \langle \nabla f(\underline{x}_k), \underline{x}_{k+1} - \underline{x}_k \rangle + \tfrac{1}{2} \langle \underline{x}_{k+1} - \underline{x}_k, G_k(\underline{x}_{k+1} - \underline{x}_k) \rangle$$

$$\geq \langle \nabla f(\underline{x}_k), \underline{x}_{k+1} - \underline{x}_k \rangle . \qquad (4.2.63)$$

The first order expansion of $\nabla f(\underline{x})$ can be written as

$$\nabla f(\underline{x}_k) = \nabla f(\underline{x}_{k-1}) + G_{k-1}(\underline{x}_k - \underline{x}_{k-1}) + \int_0^1 \{G[\underline{x}_{k-1} + t(\underline{x}_k - \underline{x}_{k+1})] - G_{k-1}\}(\underline{x}_k - \underline{x}_{k-1})dt, \qquad (4.2.64)$$

with

$$\nabla f(\underline{x}_{k-1}) + G_{k-1}(\underline{x}_k - \underline{x}_{k-1}) = \nabla \bar{q}_{k-1}(\underline{x}_k) .$$

Clearly, $\tau_k = 1$ implies $\underline{x}_p = \underline{x}_{k+1}$ and with $\alpha_k = 1$ the inequality

$$\langle \nabla \bar{q}_{k-1}(\underline{x}_k), \underline{x}_{k+1} - \underline{x}_k \rangle \geq o$$

follows from the optimality of $\underline{x}_k$ for $\bar{q}_{k-1}$. Taking the scalar product of (4.2.64) with $\underline{x}_{k+1} - \underline{x}_k$ and using (4.2.63) yields

$$- \bar{q}_k(\underline{x}_{k+1}) \leq - \int_0^1 \langle \underline{x}_{k+1} - \underline{x}_k, (G[\underline{x}_{k-1} + t(\underline{x}_k - \underline{x}_{k-1})] - G_{k-1})(\underline{x}_k - \underline{x}_{k-1}) \rangle dt$$

which, using the Schwartz inequality and

$\mu^o_{k-1} = \int_0^1 \| G[\underline{x}_{k-1} + t(\underline{x}_k - \underline{x}_{k-1})] - G_{k-1}\| dt$, becomes

$$- \bar{q}_k(\underline{x}_{k+1}) \leq \mu^o_{k-1} \|\underline{x}_{k+1} - \underline{x}_k\| \, \| \underline{x}_k - \underline{x}_{k-1}\| . \qquad (4.2.65)$$

Using (4.2.32), with $\alpha_k = 1$, and (4.2.65) results in the inequality

$$\|\underline{x}_{k+1} - \underline{x}_k\| \le \frac{2\,\mu^o_{k-1}}{\ell}\,\|\underline{x}_k - \underline{x}_{k-1}\| . \qquad (4.2.66)$$

Since $\lim_{k\to\infty} \mu^o_k = o$ and, from Theorem (4.5), $\lim_{k\to\infty} f(\underline{x}_k) = f(\underline{x}_c)$, for $\beta_k = 2\,\mu^o_{k-1}/\ell$ there exists a constant k_o such that for all $k \ge k_o$, $\beta_k < 1$. Thus

$$\begin{aligned}\|\underline{x}_c - \underline{x}_k\| &= \lim_{t\to\infty} \|\underline{x}_t - \underline{x}_k\| \\ &\le \lim_{t\to\infty} \sum_{j=k}^{t-1} \|\underline{x}_{j+1} - \underline{x}_j\| \\ &\le \beta_k \|\underline{x}_k - \underline{x}_{k-1}\| \lim_{t\to\infty} \sum_{j=k}^{t-1} \beta_k^{j-k} \\ &= \frac{\beta_k}{1-\beta_k} \|\underline{x}_k - \underline{x}_{k-1}\| \qquad (4.2.67) \\ &\le \frac{\beta_k}{1-\beta_k} (\|\underline{x}_k - \underline{x}_c\| + \|\underline{x}_c - \underline{x}_{k-1}\|)\end{aligned}$$

and rearranging yields

$$\frac{\|\underline{x}_c - \underline{x}_k\|}{\|\underline{x}_c - \underline{x}_k\| + \|\underline{x}_c - \underline{x}_{k-1}\|} \le \frac{\beta_k}{1-\beta_k} \qquad (4.2.68)$$

As $\lim_{k\to\infty} \beta_k = o$, (4.2.68) implies that

$$\lim_{k\to\infty} \frac{\|\underline{x}_c - \underline{x}_k\|}{\|\underline{x}_c - \underline{x}_{k-1}\|} = o \qquad (4.2.69)$$

which is the condition for Q-superlinear convergence, given by Ortega and Rheinboldt (1970). This proves the result for (4.2.37).

Next consider the criterion

$$f(\underline{x}_{k+1}) - f(\underline{x}_k) \leq c_1 \tau_k \langle \nabla f(\underline{x}_k), \underline{x}_p - \underline{x}_k \rangle, \; o < c_1 < 1 - \frac{\bar{\alpha}}{2} \qquad (4.2.38)$$

for choosing the step size τ_k. Using (4.2.48) and that $\mu_k \to o$ from (4.2.62) implies that for any $o < c_1 < 1 - \frac{\bar{\alpha}}{2}$, $o < \bar{\alpha} < 2$ and $k \geq K(c_1)$, the bound (4.2.50) is satisfied for $\alpha_k = \tau_k = 1$. The inequality

$$\langle \nabla f(\underline{x}_k), \underline{x}_{k+1} - \underline{x}_k \rangle \geq - \mu^o_{k-1} ||\underline{x}_{k+1} - \underline{x}_k|| \, ||\underline{x}_k - \underline{x}_{k-1}||$$

follows from (4.2.64) for $\tau_k = 1$ and $\underline{x}_k = \underline{x}_p$. The rest of the proof for (4.2.38) proceeds exactly as the previous case using (4.2.63).

Finally, convergence with (4.2.28) may be established in a similar manner as the Global Damped Newton Theorem (Ortega and Rheinboldt (1970), Theorem 14.4.3). Using the comparison (4.2.52) of the stepsize methods, the convergence of the sequence $\underline{x}_k$ to $\underline{x}_c$ follows from Theorem (4.5). Thus, for $\delta > o$, there is a sphere

$$S \triangleq \{\underline{x} \in E^n \mid ||\underline{x}_c - \underline{x}|| \leq \delta\}$$

such that for some k_o, $\underline{x}_k \in S$, $k \geq k_o$ and the iterates converge to $\underline{x}_c$ with $\tau_k = 1$. Superlinear convergence for $\tau_k = 1$ follows from the above discussion.

Quadratic convergence can be established by further assuming that

$$||G[\underline{x}] - G[\underline{y}]|| \leq M_G ||\underline{x} - \underline{y}|| \quad , \quad \forall \; \underline{x}, \underline{y} \in R$$

holds for the second derivative of f at $\underline{x}$ and $\underline{y}$. As argued above, for $k \geq K(c_1)$, $\alpha_k = \tau_k = 1$ and, for μ^o_k, $\underline{x}_p = x_k$. We can establish the bound

$$\mu_{k-1} = \int_0^1 \|G[\underline{x}_{k-1} + t(\underline{x}_k - \underline{x}_{k-1})] - G[\underline{x}_{k-1}]\| \, dt$$

$$\leq M_G \|\underline{x}_k - \underline{x}_{k-1}\| \int_0^1 t \, dt$$

$$= \frac{M_G}{2} \|\underline{x}_k - \underline{x}_{k-1}\| . \qquad (4.2.70)$$

From (4.2.68) we have

$$(1 - 2\beta_k) \|\underline{x}_c - \underline{x}_k\| \leq \beta_k \|\underline{x}_c - \underline{x}_{k-1}\| \qquad (4.2.71)$$

in which β_k is bounded by

$$\beta_k = 2\mu^o_{k-1}/\ell \leq \frac{M_G}{\ell} \|\underline{x}_k - \underline{x}_{k-1}\| \qquad (4.2.72)$$

using (4.2.70). Furthermore

$$\|\underline{x}_k - \underline{x}_{k-1}\| \leq \|\underline{x}_k - \underline{x}_c\| + \|\underline{x}_c - \underline{x}_{k-1}\|$$

and by (4.2.69) there exists $\sigma \geq o$ such that

$$\|\underline{x}_k - \underline{x}_c\| \leq \sigma \|\underline{x}_{k-1} - \underline{x}_c\| ;$$

thus

$$\|\underline{x}_k - \underline{x}_{k-1}\| \leq (1 + \sigma) \|\underline{x}_c - \underline{x}_{k-1}\| .$$

Setting $M_\sigma = (1 + \sigma)M_G/\ell$ we have

$$\beta_k \leq M_\sigma \|\underline{x}_c - \underline{x}_{k-1}\| .$$

Also since $\beta_k \to o$, there exists k_o such that $\beta_k < \frac{1}{2}$ for all $k \geq k_o$, we obtain from (4.2.71)

$$\|\underline{x}_c - \underline{x}_k\| \leq \frac{M_\sigma}{1 - 2M_\sigma\|\underline{x}_c - \underline{x}_{k-1}\|} \|\underline{x}_c - \underline{x}_{k-1}\|^2 \qquad (4.2.73)$$

which is the condition for quadratic convergence (see Ortega and Rheinboldt (1970)). □

4.2.3 Convergence Proofs: Approximate Second Derivatives

The results in Section (4.2.2) are concerned with the convergence of (4.1.7) to a global optimum assuming convex $f(\underline{x})$ and R. Convergence to a local optimum may be shown under less restricted assumptions. It will first be shown that algorithm (4.1.7) converges locally with a positive definite approximation $\hat{G}_k$ and $\bar{q}_k(\underline{x})$ replaced by

$$\hat{q}_k(\underline{x}) = \langle \nabla f(\underline{x}_k), \underline{x} - \underline{x}_k \rangle + \tfrac{1}{2} \langle \underline{x} - \underline{x}_k, \hat{G}_k(\underline{x} - \underline{x}_k) \rangle . \quad (4.2.74,a)$$

Accordingly, the strategy (4.2.26) is altered such that τ_k is chosen to satisfy

$$f(\underline{x}_{k+1}) - f(\underline{x}_k) \le c_1 \tau_k \hat{q}_k(\underline{x}_p) , \quad 0 < c_1 < 1 . \quad (4.2.74,b)$$

Consider algorithm (4.1.7). The assumption in Section (4.2.2) that G_k should be positive definite on R can be relaxed if the positive definite approximation $\hat{G}_k$ is used. Lemmas (4.5) - (4.6) hold with $\hat{G}_k, \hat{q}_k$ replacing $G_k, \bar{q}_k$. The monotonicity result in Theorem (4.3) also holds when $G_k, \bar{q}_k$ are replaced by $\hat{G}_k, \hat{q}_k$. The expression (4.2.40) can be written as

$$\begin{aligned} f(\underline{x}_{k+1}) - f(\underline{x}_k) &= \langle \nabla f(\underline{x}_k), \underline{x}_{k+1} - \underline{x}_k \rangle \\ &\quad + \tfrac{1}{2} \langle \underline{x}_{k+1} - \underline{x}_k , \hat{G}_k(\underline{x}_{k+1} - \underline{x}_k) \rangle \\ &\quad + \int_0^1 (1-t) \langle \underline{x}_{k+1} - \underline{x}_k, \{G[\underline{x}_k - t(\underline{x}_k - \underline{x}_{k+1})] - \hat{G}_k\} (\underline{x}_{k+1} - \underline{x}_k) \rangle dt \end{aligned}$$

(4.2.75,a)

$$\leq \hat{q}_k(\underline{x}_{k+1}) + \tau_k^2 \hat{\mu}_k \|\underline{x}_p - \underline{x}_k\|^2$$

$$\leq \tau_k \hat{q}_k(\underline{x}_p) + \tau_k^2 \hat{\mu}_k \|\underline{x}_p - \underline{x}_k\|^2$$

where $\hat{\mu}_k$ given by

$$\hat{\mu}_k = \int_0^1 (1 - t) \|G[\underline{x}_k - \tau_k t(\underline{x}_k - \underline{x}_p)] - \hat{G}_k\| \, dt \qquad (4.2.75,b)$$

and the bound $\hat{\mu}_k < \infty$, $\forall \underline{x}_k \in R$, replacing the bound on G_k in (4.2.34), are the only modifications required. For $\hat{G}_k$ replacing G_k in (4.2.45), it is assumed that the choice of $\hat{G}_k$ provides the bound $\hat{\mu}_k \leq L < \infty$, for L not related to (4.2.34). Thus, inequality (4.2.46) can be rewritten with $\hat{G}_k$ replacing G_k hence establishing the monotonicity results for (4.2.37), with $\hat{q}_k(\underline{x}_p)$ replacing $\bar{q}_k(\underline{x}_p)$, and for (4.2.38). Theorem (4.4) is also valid when G_k, $\bar{q}_k$ are replaced by $\hat{G}_k$, $\hat{q}_k$. The only change arises from relaxing the assumption (4.2.34) on G_k. Since $\hat{G}_k$ replaces G_k in (4.2.34), the assumption that $f(\underline{x})$ is bounded from below on R has to be made explicitly. The proof of Theorem (4.4) applies to $\hat{G}_k$, $\hat{q}_k$. Finally, Lemma (4.7) also holds for $\hat{G}_k$, $\hat{q}_k$ replacing G_k, $\bar{q}_k$ without further modification to its proof.

The convexity assumptions on f and R are only used explicitly in Theorem (4.5) in establishing $\underline{x}_c$ as the global optimum of f on R. An alternative proof of the convergence of $\{\underline{x}_k\}$ to $\underline{x}_c$, without the convexity assumption on f, will be given in Theorem (4.8) below.

The projection problem (4.1.12) in algorithm (4.1.7) may be written as

$$\underline{x}_p = \arg\min \ \{ \tfrac{1}{2} \| \underline{x} - \underline{x}_k + \alpha_k \hat{H}_k \nabla f(\underline{x}_k) \|^2_{\hat{G}_k} \ | \ \underline{x} \in R \} \qquad (4.2.76)$$

where α_k is chosen to satisfy

$$o < \alpha_k \le \bar{\alpha} < 2 \qquad (4.2.77)$$

and

$$\underline{x}_{k+1} = \underline{x}_k + \tau_k (\underline{x}_p - \underline{x}_k) \in R \qquad (4.2.78)$$

for

$$o < \tau_k \le 1 .$$

If there is more than one solution to the minimisation problem in (4.2.76), then $\underline{x}_p$ is set to the solution closest to $\underline{x}_k$, i.e.

$$\|\underline{x}' - \underline{x}_k\| \ge \|\underline{x}_p - \underline{x}_k\| \qquad (4.2.79)$$

where $\underline{x}'$ denotes any other solution of the minimisation in (4.2.76). It is shown below that the sequence $\{\underline{x}_k\}$ generated by algorithm (4.1.7) converges to $\underline{x}_c$, the constrained minimum of $f(\underline{x})$ on R, if $\{\underline{x}_k\}$ satisfies either $\lim_{k\to\infty} \hat{q}_k(\underline{x}_p) = o$ or (4.2.56). The condition $\lim_{k\to\infty} \hat{q}_k(\underline{x}_p) = o$ can be interpreted as (4.2.55) with $\hat{q}_k(\underline{x}_p)$ replacing $\bar{q}_k(\underline{x}_p)$.

For linear inequality constrained R and $\hat{G}_k$ positive definite, (4.2.76) is a positive definite quadratic programming problem. Clearly $\underline{x}_p$ is unique and hence the global minimum (see Chapter 3). The problem of nonunique solutions for the minimisation in (4.2.76) arises when $\hat{G}_k$ is not strictly positive definite. For positive semidefinite $\hat{G}_k$ the unconstrained descent direction $- \hat{H}_k \nabla f(\underline{x}_k)$ may be replaced by the direction $\underline{d}$ satisfying

$$\hat{G}_k \underline{d} = - \nabla f(\underline{x}_k)$$

and the quadratic minimisation problem in (4.2.76) may be written as

$$\underline{x}_p = \arg\min \{ \tfrac{1}{2} \|\underline{x} - \underline{x}_k - \alpha_k \underline{d}\|^2_{\hat{G}_k} \mid \underline{x} \in R \} \qquad (4.2.80)$$

and solved by a quadratic programming method that does not require G_k to be positive definite (see, e.g. Gill and Murray (1978). For indefinite $\hat{G}_k$, the eigenvalues of $\hat{G}_k$ may be negative, zero and/or positive. Ignoring its constant term, the quadratic minimsation in (4.2.76) may be written as

$$\underline{x}_p = \arg\min \{ \alpha_k \langle \nabla f(\underline{x}_k), \underline{x} - \underline{x}_k \rangle + \tfrac{1}{2} \langle \underline{x} - \underline{x}_k, \hat{G}_k(\underline{x} - \underline{x}_k) \rangle \mid \underline{x} \in R \}$$

(4.2.81)

which can be solved using the indefinite quadratic programming algorithm due to Gill and Murray (1978). When the quadratic minimisation problem does not have a unique solution, (4.2.79) implies that $\underline{x}_p$ should be chosen as the solution nearest, in Euclidean norm, to $\underline{x}_k$.

When the solution of the quadratic minimisation problem is nonunique, $\underline{x}_p$ cannot be interpreted as the projection of $\underline{x}_k - \alpha_k \hat{H}_k \nabla f(\underline{x}_k)$ onto R . This is due to the uniqueness property of projections (Luenberger (1969)). This property requires $\underline{x}_p$ to be the unique solution of the minimisation in (4.2.76).

In the subsequent analysis $\hat{G}_k$ is assumed to be positive definite. Hence, $\underline{x}_p$ will be taken as the unique solution in (4.2.76). The objective function will be assumed to be twice continuously differentiable, possessing finite first and second derivatives. To establish superlinear convergence, it will further be assumed that at the optimum the projection of the Hessian of f is positive definite on the tangent hyperplanes of the active constraints. □

To start the discussion on the use of $\hat{G}_k$, the extensions of earlier results to $\hat{G}_k$ are collectively stated in the following theorem for convex R.

Theorem (4.7)

If G_k, $\bar{q}_k(\underline{x}_p)$, and subproblem (4.1.12) are replaced by $\hat{G}_k$, $\hat{q}_k(\underline{x}_p)$ and subproblem (4.2.76) respectively, then Lemmas (4.2), (4.3), (4.5), (4.6), (4.7), Theorems (4.3), (4.4) hold for $\{\underline{x}_k\}$ generated by algorithm (4.1.7) with either one of the stepsize strategies (4.2.38) and (4.2.74,b).

Proof

If $\underline{x}_p$ is taken as the projection of $\underline{x}_k - \alpha_k \hat{H}_k \nabla f(\underline{x}_k)$ onto R, defined for positive definite $\hat{G}_k$ by (4.2.76), it satifies the requirements of Lemmas (4.2), (4.3). Hence these Lemmas hold for $\hat{G}_k$ and subproblem (4.2.76).

Lemma (4.5) holds for any positive semidefinite matrix $\hat{G}_k$. Thus (4.2.25) may be rewritten as

$$\hat{q}_k(\underline{x}_{k+1}) \leq \tau_k \hat{q}_k(\underline{x}_p) , \quad 0 \leq \tau_k \leq 1 . \tag{4.2.82}$$

Lemma (4.6) holds for any positive definite $\hat{G}_k$. Thus (4.2.29) may be written as

$$q_k(\underline{x}_p) \leq \left(\frac{1}{2} - \frac{1}{\alpha_k} \right) \|\underline{x}_p - \underline{x}_k\|^2_{\hat{G}_k} \leq 0 , \quad 0 < \alpha_k \leq 2 . \tag{4.2.83}$$

Theorems (4.3), (4.4) and Lemma (4.7) use the information that $\underline{x}_p - \underline{x}_k$ is a descent direction. Since Lemma (4.3) is valid for $\hat{G}_k$, (4.2.6) can be written as

$$o \geq -\|\underline{x}_p - \underline{x}_k\|^2_{\hat{G}_k} \geq \alpha_k < \nabla f(\underline{x}_k), \underline{x}_p - \underline{x}_k > . \qquad (4.2.84)$$

From (4.2.83) it follows that $\hat{q}_k(\underline{x}_p) \leq o$ and from (4.2.84) it follows that $\underline{x}_p - \underline{x}_k$ is a descent direction. Hence the proposed steplength strategies (4.2.74,b) and (4.2.38) are valid. By replacing G_k, H_k with $\hat{G}_k$, $\hat{H}_k$ in (4.2.34) and (4.2.35), assuming that $f(\underline{x})$ possesses finite first and second derivatives, using the second order expansion (4.2.75) with (4.2.82) instead of (4.2.39) - (4.2.40), the monotonic decrease of $f(\underline{x}_k)$ for $\{\underline{x}_k\}$ is established in the same manner as in the proof of Theorem (4.3).

If G_k is replaced by $\hat{G}_k$ in (4.2.34) and f is restricted to be bounded from below on R, Theorem (4.4) can be restated for $\hat{G}$ Thus, for $\hat{G}_k$, this requires (4.2.56) to hold and (4.2.55) to be replaced by

$$\lim_{k\to\infty} \hat{q}_k(\underline{x}_p) = o \, . \qquad (4.2.85)$$

As the above monotonicity result for $f(\underline{x}_k)$ensures (4.2.57), using (4.2.83), (4.2.84) and the above extensions of Lemmas (4.3), (4.6), the required result is established in the same manner as in Theorem (4.4).

For $\hat{G}_k$ and (4.2.85) replacing G_k and (4.2.55) respectively, Lemma (4.7) can be restated for $\hat{G}$. Similar to the original proof of Lemma (4.7), for condition (4.2.85) inequality (4.2.83) provides the result. For (4.2.56) inequality (4.2.84), as the extension of Lemma (4.3), provides the result. □

The following two definitions from Ortega and Rheinboldt (1970) state the properties assumed in the theorem below.

Definition (4.3)

A functional $f : D \subset E^n \to E^1$ is hemivariate on a set $D_o \subset D$ if it is not constant on any line segment of D_o, i.e. $\nexists$ distinct points $\underline{x}, \underline{y} \in D_o$ such that $(1-t)\underline{x} + t\underline{y} \in D_o$ and $f((1-t)\underline{x} + t\,\underline{y}) = f(\underline{x}), \ \forall \ t \in [o, 1]$.

Definition (4.4)

Given $f : D \subset E^n \to E^1$, a sequence $\{\underline{x}_k\}$ in some subset $D_o \subset D$ is strongly downward in D_o if $(1-t)\underline{x}_k + t\,\underline{x}_{k+1} \in D_o$, $\forall\ t \in [o,1]$, and $f(\underline{x}_k) \geq f((1-t)\underline{x}_k + t\,\underline{x}_{k+1}) \geq f(\underline{x}_{k+1})$, $\forall\ t \in [o,1]$.

Theorem (4.8)

Assume that $f : D \subset E^n \to E^1$ is continuously differentiable on the open set D and that there is an $\underline{x}_o \in D$ such that the level set

$$L^o = \{\underline{x} \in D \mid f(\underline{x}) \leq f(\underline{x}_o)\} \tag{4.2.86}$$

is compact and $L^o \cap R \neq \phi$. Assume that f is hemivariate and has finitely many points satisfying the necessary conditions of optimality over $L^o \cap R$. Let the sequence $\{\underline{x}_k\}$ defined by (4.2.76), for positive definite $\hat{G}_k$, and (4.2.1) be strongly downward and satisfy either $\lim_{k\to\infty} \langle \nabla f(\underline{x}_k), \underline{x}_p - \underline{x}_k \rangle = o$ or $\lim_{k\to\infty} \hat{q}_k(\underline{x}_p) = o$. Then $\{\underline{x}_k\}$ converges to one of the points satisfying the necessary conditions of optimality of $f(\underline{x})$ over $L^o \cap R$.

Proof

From (4.2.29), which holds for $\hat{G}_k$ as well as G_k, and (4.2.74), the inequality

$$o \geq \hat{q}_k(\underline{x}_p) \geq \langle \nabla f(\underline{x}_k), \underline{x}_p - \underline{x}_k \rangle \tag{4.2.87}$$

follows. Thus, $\lim_{k\to\infty} \langle \nabla f(\underline{x}_k), \underline{x}_p - \underline{x}_k \rangle = o$ implies $\lim_{k\to\infty} \hat{q}_k(\underline{x}_p) = o$. The converse is also true since with (4.2.83) $\lim_{k\to\infty} \hat{q}_k(\underline{x}_p) = o$ implies $\lim_{k\to\infty} \|\underline{x}_p - \underline{x}_k\| = o$. The inequalities $\|\nabla f(\underline{x}_k)\| \, \|\underline{x}_p - \underline{x}_k\| \geq -\langle \nabla f(\underline{x}_k), \underline{x}_p - \underline{x}_k \rangle \geq \alpha_k \|\underline{x}_p - \underline{x}_k\|^2_{\hat{G}_k}$ are obtained using the Schwartz inequality and (4.2.84). Hence $\lim_{k\to\infty} \hat{q}_k(\underline{x}_p) = o$ implies $\lim_{k\to\infty} \langle \nabla f(\underline{x}_k), \underline{x}_p - \underline{x}_k \rangle = o$. Thus, only $\lim_{k\to\infty} \langle \nabla f(\underline{x}_k), \underline{x}_p - \underline{x}_k \rangle = o$ will be considered.

In the limit, the projection problem (4.2.76) can be written as,

$$\begin{aligned}
&\lim_{k\to\infty} \|\underline{x}_p - \underline{x}_k + \alpha_k \hat{H}_k \nabla f(\underline{x}_k)\|^2_{\hat{G}_k} \\
&= \lim_{k\to\infty} \min \{ \|\underline{x} - \underline{x}_k + \alpha_k \hat{H}_k \nabla f(\underline{x}_k)\|^2_{\hat{G}_k} \mid \underline{x} \in R \} \\
&= \lim_{k\to\infty} \min \{ \|\underline{x} - \underline{x}_k\|^2_{\hat{G}_k} + 2\alpha_k \langle \nabla f(\underline{x}_k), \underline{x} - \underline{x}_k \rangle \\
&\qquad + \alpha_k^2 \|\hat{H}_k \nabla f(\underline{x}_k)\|^2_{\hat{G}_k} \mid \underline{x} \in R \} \\
&= \lim_{k\to\infty} \alpha_k^2 \|\hat{H}_k \nabla f(\underline{x}_k)\|^2_{\hat{G}_k} .
\end{aligned} \tag{4.2.88}$$

It follows from (4.2.88) that

$$\lim_{k\to\infty} \|\underline{x}_p - \underline{x}_k\|^2_{\hat{G}_k} = o . \tag{4.2.89}$$

The Lagrange multipliers associated with the minimisation in (4.2.76) at $\underline{x}_p$ are clearly nonnegative. Furthermore, since $\underline{x}_p, \underline{x}_k \in R$, the inequality

$$\underline{g}(\underline{x}_k) \geq o \quad , \forall k, \tag{4.2.90}$$

follows. From (4.2.88) it follows that the projection of $\lim_{k\to\infty} - \alpha_k \hat{H}_k \nabla f(\underline{x}_k)$ on R vanishes . This implies that $\lim_{k\to\infty} \nabla f(\underline{x}_k)$ is a linear combination of the constraint tangent normals at $\lim_{k\to\infty} \underline{x}_k$ and since by (4.2.90) $\lim_{k\to\infty} \underline{x}_k$ is feasible on R , $\lim_{k\to\infty} \underline{x}_k$ satisfies the necessary conditions of optimality. Also, it follows from (4.2.89) that the projection (i.e. the closest solution to $\underline{x}_k$ in (4.2.76)) of $\lim_{k\to\infty} \underline{x}_k - \alpha_k \hat{H}_k \nabla f(\underline{x}_k)$ on R is the same as $\lim_{k\to\infty} x_k$.

Since it follows from (4.2.90) that

$$\lim_{k\to\infty} x_k \in R \tag{4.2.91}$$

this statement is equivalent to the above necessary conditions for optimality (Levitin and Polyak (1966)).

Because of the compactness of L^o , $\{\underline{x}_k\}$ has convergent subsequences and if $\lim_{i\to\infty} \underline{x}_{k_i} = \underline{x}_c$ then, by continuity of ∇f , $\underline{x}_c$ satisfies $L^o \cap R$ and $<\nabla f(\underline{x}_c), \underline{x}_p - \underline{x}_c> = o$ where $\underline{x}_p$ is the projection of $\underline{x}_c - \hat{H}_c \nabla f(\underline{x}_c)$ onto R . Let

$$\delta_k = \inf \{ ||\underline{x}_k - \underline{x}|| \mid \underline{x} \in L^o \cap R , <\nabla f(\underline{x}), \underline{x}_p - \underline{x}> = o \}$$

where $\underline{x}_p$ is the projection of $\underline{x} - \alpha_x \hat{H}_x \nabla f(\underline{x})$ with $o < \alpha_x \leq \bar{\alpha} < 2$ and $\hat{H}_x$ a positive definite approximation to the inverse Hessian at $\underline{x}$. Suppose that $\lim_{i\to\infty} \delta_{k_i} = \delta$. Then $\underline{x}_{k_i}$ has a convergent subsequence, and, since this subsequence must have a limit point in $\{\underline{x} \in L^o \cap R \mid <\nabla f(\underline{x}) , \underline{x} - \underline{x}_p> = o \}$, it follows that $\delta = o$. Thus

$$\lim_{k\to\infty} \inf \{ ||\underline{x}_k - \underline{x}_c|| \mid \underline{x}_c \in L^o \cap R , < \nabla f(\underline{x}_c), \underline{x}_c - \underline{x}_p > = o \} = o$$

where $\underline{x}_p$ above denotes the projection of $\underline{x}_c - \alpha H_c \nabla f(\underline{x}_c)$ onto R .

Hence when $\underline{x}_c$ is unique in $L^o \cap R$, the sequence $\{\underline{x}_k\}$ converges to it. If, however, $\{\underline{x} \in L^o \cap R \mid \langle \nabla f(\underline{x}), \underline{x} - \underline{x}_p \rangle = o\}$ consists of finitely many points, convergence to one of these points is assured if

$$\lim_{k\to\infty} (\underline{x}_{k+1} - \underline{x}_k) = \underline{o} \tag{4.2.92}$$

holds. This condition is satisfied for any strongly downward sequence provided f is hemivariate (see, Ortega and Rheinboldt (1970, Theorem 14.1.3)). Thus, if a sequence $\{x_k\} \subset L^o$ satisfies this condition and $\lim_{k\to\infty} \langle \nabla f(\underline{x}_k), \underline{x}_k - \underline{x}_p \rangle = o$, it may be shown to have a single limit point in $\{\underline{x} \in L^o \cap R \mid \langle \nabla f(\underline{x}), \underline{x} - \underline{x}_p \rangle = o\}$. This is done by restating a result in unconstrained optimisation (Ortega and Rheinboldt (1970, Theorem 14.1.5)) for constrained optimisation.

Let Λ be the set of limit points of $\{\underline{x}_k\}$. As in the discussion above for $\underline{x}_c$, any limit point is also a point satisfying the necessary conditions for optimality. Thus

$$\Lambda \subset \{ \underline{x} \in L^o \cap R \mid \langle \nabla f(\underline{x}), \underline{x} - \underline{x}_p \rangle = o \}$$

and Λ is finite. Suppose $\Lambda = \{\underline{x}_c^1, \ldots, \underline{x}_c^m\}$ i.e. $\{\underline{x}_k\}$ has $m > 1$ limit points. Then

$$\delta = \min \{ \|\underline{x}_c^i - \underline{x}_c^j\| \mid i \neq j,\ i, j = 1, \ldots m \} > o$$

and $k_o \geq o$ can be chosen such that

$$\underline{x}_k \in \bigcup_{i=1}^{m} \{ \underline{x} \in L^o \mid \|\underline{x} - \underline{x}_c^i\| < \delta/4 \} \quad \text{and}$$

$\|\underline{x}_k - \underline{x}_{k+1}\| \leq \delta/4$, $\forall\, k \geq k_o$. However,

$\underline{x}_{k_1} \in \{ \underline{x} \in L^o \mid \|\underline{x} - \underline{x}_c^1\| < \delta/4 \}$, for some $k_1 \geq k_o$,

implies that

$$\|\underline{x}_c^i - \underline{x}_{k_1+1}\| \geq \|\underline{x}_c^i - \underline{x}_c^1\| - (\|\underline{x}_c^1 - \underline{x}_{k_1}\| + \|\underline{x}_{k_1} - \underline{x}_{k_1+1}\|)$$

$$\geq \delta - 2\,\delta/4 = \delta/2, \quad i \geq 2 ,$$

and hence $\underline{x}_{k_1+1} \in \{ \underline{x} \in L^o \mid \|\underline{x}_c^1 - \underline{x}\| < \delta/4 \}$. By induction, $\underline{x}_k \in \{ \underline{x} \in L^o \mid \|\underline{x}_c^1 - \underline{x}\| < \delta/4 \} \;\forall\, k \geq k_1$, thus $\underline{x}_c^2, \ldots, \underline{x}_c^m$ cannot be limit points of $\{\underline{x}_k\}$; hence $m = 1$, which is the required result. □

Theorem (4.7) establishes the condition

$$\lim_{k\to\infty} \|\underline{x}_k - \underline{x}_p\| = o$$

for algorithm (4.1.7) implemented with a positive definite $\hat{G}_k$ and either one of the strategies (4.2.38) or (4.2.74,b) for τ_k . Using $\underline{x}_{k+1} - \underline{x}_k = \tau_k(\underline{x}_p - \underline{x}_k)$, $o < \tau_k \leq 1$, it follows that

$$\lim_{k\to\infty} \|\underline{x}_{k+1} - \underline{x}_k\| = \lim_{k\to\infty} \tau_k \|\underline{x}_p - \underline{x}_k\| = o .$$

Thus, the restriction in Theorem (4.8) that f should be hemivariate and $\{\underline{x}_k\}$ strongly downward (to ensure $\lim_{k\to\infty} (\underline{x}_{k+1} - \underline{x}_k) = \underline{o}$) are not explicitly required for (4.1.7). However, an important assumption in this case is that μ_k in (4.2.42) or $\hat{\mu}_k$ in (4.2.75,b) are bounded from above. □

Theorem (4.8) also applies to alternative stepsize mechanisms which set $\tau_k = 1$ in (4.2.1) and adjust α_k in (4.1.12) to satisfy (4.2.85) or (4.2.56). Such a strategy could be used in the Goldstein-Levitin-Polyak algorithm (see Sections (1.2.1),(4.1.1)).

The stepsize strategies (4.2.38) and

$$f(\underline{x}_k) - f(\underline{x}_{k+1}) \geq - c_1 \tau_k \hat{q}_k(\underline{x}_p) \qquad (4.2.93)$$

can be expressed in terms of $\underline{x}_p - \underline{x}_k$. Using (4.2.83),(4.2.93) can be written as

$$f(\underline{x}_k) - f(\underline{x}_{k+1}) \geq - c_1 \tau_k \left(\frac{1}{2} - \frac{1}{\alpha_k}\right) \|\underline{x}_p - \underline{x}_k\|^2_{\hat{G}_k} \qquad (4.2.94)$$

and using (4.2.84),(4.2.38) can be written as

$$f(\underline{x}_k) - f(\underline{x}_{k+1}) \geq \frac{c_1 \tau_k}{\alpha_k} \|\underline{x}_p - \underline{x}_k\|^2_{\hat{G}_k} \qquad (4.2.95)$$

for $0 < \alpha \leq \bar{\alpha} < 2$. Using the first order optimality conditions of the minimisation in (4.2.76), an alternative expression for $\underline{x}_p - \underline{x}_k$ can be obtained

$$\hat{G}_k (\underline{x}_p - \underline{x}_k) = - \alpha_k(\nabla f(\underline{x}_k) - N_p \underline{\lambda}_p / \alpha_k) \qquad (4.2.96)$$

where the columns of N_p denote the linearly independent constraint normals evaluated at $\underline{x}_p$ and $\underline{\lambda}_p$ is the vector of Lagrange multipliers associated with the quadratic minimisation (4.2.76) and $\underline{\lambda}_{k+1}$ is defined as

$$\underline{\lambda}_{k+1} = \underline{\lambda}_p / \alpha_k . \qquad (4.2.97)$$

The following lemma shows that $\underline{\lambda}_k$ converges to $\underline{\lambda}_c$, the Lagrange multiplier vector at $\underline{x}_c$.

<u>Lemma (4.8)</u>

Let $\underline{x}_p$ be computed by (4.2.76) and $\underline{\lambda}_k$ given by (4.2.97). If the sequence $\{\underline{x}_k\}$ converges to $\underline{x}_c$ such that $\lim_{k\to\infty} \|\underline{x}_p - \underline{x}_k\| = 0$ then $\lim_{k\to\infty} \underline{\lambda}_k = \underline{\lambda}_c$ and for sufficiently large k, $\underline{\lambda}_k$ predicts the constraints active at $\underline{x}_c$.

<u>Proof</u>

Since $\{\underline{x}_k\}$ converges to $\underline{x}_c$ and $\lim_{k\to\infty} \|\underline{x}_p - \underline{x}_k\| = 0$, $\lim_{k\to\infty} \underline{\lambda}_k = \underline{\lambda}_c$ is immediately established by taking limits of both sides of (4.2.96) and using (4.2.97). Thus,

$$\lim_{k\to\infty} N_p \underline{\lambda}_p/\alpha_k = \lim_{k\to\infty} \nabla f(\underline{x}_k) = \nabla f(\underline{x}_c) = N_c \underline{\lambda}_c \tag{4.2.98}$$

holds where N_c is the matrix of linearly independent active constraint normals at $\underline{x}_c$. As N_p, N_c are full rank $\lim_{k\to\infty} \underline{\lambda}_p/\alpha_k = \underline{\lambda}_c$ follows. Since $\underline{x}_k \in R$, $\underline{g}(\underline{x}_k) \geq \underline{o}$ and for k sufficiently large the inequality

$$g_i(\underline{x}_k) > o$$

holds for the ith inactive constraint at $\underline{x}_c$ provided strict complementarity holds. Thus, for k sufficiently large, none of the inactive constraints are predicted to be active at $\underline{x}_c$. The required result follows from the strict compliment-arity condition at $\underline{x}_c$ (i.e. $g_i(\underline{x}_c) > o$ implies $\lambda_c^i = o$, $g_j(\underline{x}_c) = o$ implies $\lambda_c^j > o$; for λ^i, λ^j denoting elements of $\underline{\lambda}_c$ corresponding to g_i and g_j respectively). □

A consequence of Lemma (4.8) is that for sufficiently large k, the inactive constraints at $\underline{x}_c$ do not affect the computation of $\underline{x}_k$. At that stage it would make no difference if the active constraints at $\underline{x}_c$ were treated as equality constraints. This leads to the existence of a neighbourhood of $\underline{x}_c$ given by

$$S = \{ \underline{x} \in E^n \mid ||\underline{x} - \underline{x}_c|| < \delta \} \tag{4.2.99}$$

with $\delta > o$ chosen such that the constraints active at $\underline{x}_c$ are also those that are active at $\underline{x}_k$ if $\underline{x}_k \in S$. Hence for k large enough $g_i(\underline{x}_c) = o$ implies $g_i(\underline{x}_k) = o$, and $g_i(\underline{x}_c) > o$ implies $g_i(\underline{x}_k) > o$ for $\underline{x}_k \in S$. If $\underline{g}^c$ denotes the vector of active constraints at $\underline{x}_c$, then

$$\underline{g}^c(\underline{x}_k) = \underline{o} \quad , \quad \forall\ \underline{x}_k \in S$$

holds for $\underline{x}_p$ given by (4.2.76). Thus the Lagrange multipliers associated with (4.2.76) are such that $\lambda_c^j > o$ implies $\lambda_k^j > o$ and $\lambda_c^i = o$ implies $\lambda_p^i = o$. The existence of S with $\delta > o$ has also been considered and invoked by Han (1976), Psenichny and Danilin(1978, Lemma 5.2) .

Using the mean value theorem (Ortega and Rheinboldt (1970, Theorem 3.2.2)) the active constraints at $\underline{x}_k$ may be expressed as

$$\underline{g}^c(\underline{x}_{k+1}) - \underline{g}^c(\underline{x}_k) = \underline{o} = N^T(\underline{x}_{k+1},\underline{x}_k)\ (\underline{x}_{k+1} - \underline{x}_k)$$

for $\underline{x}_k,\underline{x}_{k+1} \in R \cap S$ where

$$N(\underline{x}_{k+1},\underline{x}_k) = [\nabla g_1^c(\underline{x}_k + t_1(\underline{x}_{k+1} - \underline{x}_k)),\ldots, \nabla g_m^c(\underline{x}_k + t_m(\underline{x}_{k+1} - \underline{x}_k))]$$

$t_1, \ldots, t_m \in (0,1)$ and m is the number of active constraints. If the linearly independent columns of $N(\underline{x}_{k+1},\underline{x}_k)$ form the columns of N then the projection operator $\bar{P}_k$ is defined by

$$\bar{P}_k = I - N(N^TN)^{-1}N^T . \tag{4.2.100}$$

Clearly $\bar{P}_k\bar{P}_k = \bar{P}_k$ and $\bar{P}_k(\underline{x}_{k+1} - \underline{x}_k) = \underline{x}_{k+1} - \underline{x}_k$. As $\{\underline{x}_k\}$ converges to $\underline{x}_c$ and $\lim_{k\to\infty} \|\underline{x}_{k+1} - \underline{x}_k\| = o$, $N(\underline{x}_{k+1},\underline{x}_k)$ converges to the active constraint normals at $\underline{x}_c$ and $\bar{P}_k$ becomes the operator projecting vectors in E^n onto the active constraint normals at $\underline{x}_c$. $\bar{P}_k$ will be used in Theorem (4.9) below to characterise the projection of G_c and $\hat{G}_k$ onto the active constraints.

<u>Theorem (4.9)</u>

Let $f : E^n \to E^1$ be differentiable in the open set $D \subset E^n$ and assume that for some $\underline{x}_c$ in D,G_c, the Hessian of f at $\underline{x}_c$, is

Lipschitz continuous. Let $\{\underline{x}_k\}$ be defined by the following:

$$\underline{x}_u = \underline{x}_k - \alpha_k \hat{H}_k \nabla f(\underline{x}_k) , \quad 0 < \alpha_k \le \bar{\alpha} < 2 \tag{4.2.101}$$

$$\underline{x}_p = \arg\min \{ \tfrac{1}{2} \|\underline{x} - \underline{x}_u\|^2_{\hat{G}_k} \mid \underline{x} \in R \} \tag{4.2.102}$$

where R is the feasible region and $\hat{H}_k = \hat{G}_k^{-1}$ is symmetric positive definite with $\ell \|\underline{v}\|^2 \le \langle \underline{v}, \hat{G}_k \underline{v} \rangle$, $\ell > 0$, $\forall \underline{v} \in E^n$, and let $0 < \tau_k \le 1$ be selected as the largest number for which either

$$f(\underline{x}_{k+1}) - f(\underline{x}_k) \le \tau_k c_1 \hat{q}_k(\underline{x}_p), \quad 0 < c_1 < 1 \tag{4.2.103}$$

holds, or as an alternative strategy

$$f(\underline{x}_{k+1}) - f(\underline{x}_k) \le \tau_k c_1 \langle \nabla f(\underline{x}_k), \underline{x}_p - \underline{x}_k \rangle, \quad 0 < c_1 < 1 - \frac{\bar{\alpha}}{2} , \tag{4.2.104}$$

holds where $\underline{x}_{k+1}$ is given by

$$\underline{x}_{k+1} = \underline{x}_k + \tau_k(\underline{x}_p - \underline{x}_k) .$$

Suppose $\{\underline{x}_k\} \subset D$ and converges to $\underline{x}_c$. There exist two positive numbers t_1, t_2, an integer $K_o \ge 0$ such that if

$$\|\underline{x}_{K_o} - \underline{x}_c\| \le t_1 \tag{4.2.105,a}$$

$$\|\hat{G}_k - G_c\| \le t_2 , \quad k \ge K_o \tag{4.2.105,b}$$

then for $k \ge K_o$, $\alpha_k = \tau_k = 1$ and $\{\underline{x}_k\}$ is R-linearly convergent. Furthermore, if for $\{\underline{x}_k\} \subset S$

$$\lim_{k\to\infty} \frac{\| \bar{P}_{k+1} (\hat{G}_k - G_c) \bar{P}_k (\underline{x}_{k+1} - \underline{x}_k) \|}{\|\underline{x}_{k+1} - \underline{x}_k\|} = 0 \tag{4.2.106}$$

holds, then $\{\underline{x}_k\}$ is Q-superlinearly convergent.

Proof

Consider first the stepsize strategy (4.2.103). Using (4.2.82) and the second order expansion of $f(\underline{x})$ as in (4.2.75) with positive definite $\hat{G}_k$ yields

$$f(\underline{x}_{k+1}) - f(\underline{x}_k) \le \tau_k \hat{q}_k(\underline{x}_p) + \int_0^1 (1-t) < \underline{x}_{k+1} - \underline{x}_k , \{G[\underline{x}_k - t(\underline{x}_k - \underline{x}_{k+1})] - G_c + G_c - \hat{G}_k\} (\underline{x}_{k+1} - \underline{x}_k) > dt \tag{4.2.107}$$

$$\le \tau_k \hat{q}_k(\underline{x}_p) \left(1 - \frac{\tau_k}{\ell\xi} \left[\tfrac{1}{2} \frac{\|(G_c - \hat{G}_k)(\underline{x}_{k+1} - \underline{x}_k)\|}{\|\underline{x}_{k+1} - \underline{x}_k\|} + \mu_k^c \right] \right)$$

$$\le \tau_k \hat{q}_k(\underline{x}_p) \left(1 - \frac{\tau_k}{\ell\xi} \left[\tfrac{1}{2} \|G_c - \hat{G}_k\| + \mu_k^c \right] \right)$$

where

$$\mu_k^c = \int_0^1 \| G[\underline{x}_k - t(\underline{x}_k - \underline{x}_{k+1})] - G_c \| \, dt$$

and $\xi = -(\frac{1}{2} - \frac{1}{\alpha_k})$. Thus, as in (4.2.44), there exist $0 < \tau_k \le 1$ such that

$$0 < c_1 \le 1 - \frac{\tau_k}{\ell\xi} \left[\tfrac{1}{2} \|G_c - \hat{G}_k\| + \mu_k^c \right] < 1 . \tag{4.2.109}$$

As $\lim_{k\to\infty} \underline{x}_k \to \underline{x}_c$, by continuity $\mu_k^c \to 0$ as $k \to \infty$. Hence for some $t > 0$ and $K_0 \ge 0$ and $\underline{x}_k , k \ge K_0$, satisfying $\underline{x}_k \in \{\underline{x} \in E^n \mid \|\underline{x} - \underline{x}_c\| < t\}$ such that $\ell\xi - \mu_k^c > 0$, the inequality

$$\tfrac{1}{2} \|G_c - \hat{G}_k\| + \mu_k^c < \ell\xi$$

holds. Thus,

$$\|G_c - \hat{G}_k\| < 2(\ell\xi - \mu_k^c)$$

and τ_k can be chosen to be unity provided there exist $\hat{G}_k$, positive definite and chosen to be close enough to G_c such that the above inequality is satisfied. Clearly, if G_c is positive definite, then there exist positive definite matrices, arbitrarily close to G_c, such that (4.2.105,b) is satisfied. Furthermore, $\alpha_k = 1$ yields $\xi = \frac{1}{2}$ and for some $t_1 > o$ and $\underline{x}_k$ satisfying $\underline{x}_k \in \{\underline{x} \in E^n \mid \|\underline{x} - \underline{x}_c\| < t_1\}$ such that $\frac{\ell}{2} - \mu_k^c > o$, the inequality

$$\|G_c - \hat{G}_k\| \leq t_2 < 2(\frac{\ell}{2} - \mu_k^c) \tag{4.2.110}$$

is to be satisfied by all $\hat{G}_k$. Again, if G_c is positive definite, then positive definite $\hat{G}_k$ can always be chosen close enough to G_c to satisfy (4.2.110) .

Thus for $\|\hat{G}_k - G_c\|$ satisfying the above inequality, the steplengths $\tau_k = \alpha_k = 1$ follow. The linear convergence rate is established below using a reasoning similar to that in Theorem (4.6).

The inequality

$$\begin{aligned} \hat{q}_k(\underline{x}_{k+1}) &= \langle \nabla f(\underline{x}_k), \underline{x}_{k+1} - \underline{x}_k \rangle + \tfrac{1}{2}\langle \underline{x}_{k+1} - \underline{x}_k, \hat{G}_k(\underline{x}_{k+1} - \underline{x}_k)\rangle \\ &\geq \langle \nabla f(\underline{x}_k), \underline{x}_{k+1} - \underline{x}_k \rangle \end{aligned} \tag{4.2.111}$$

follows since $\hat{G}_k$ is positive definite. The first order expansion of $\nabla f(\underline{x})$ can be written as

$$\begin{aligned} \nabla f(\underline{x}_k) = \nabla f(\underline{x}_{k-1}) &+ \hat{G}_{k-1}(\underline{x}_k - \underline{x}_{k-1}) \\ &+ \int_0^1 \| G[\underline{x}_{k-1} + t(\underline{x}_k - \underline{x}_{k-1})] - G_c + G_c - \hat{G}_{k-1} \} \\ &\qquad (\underline{x}_k - \underline{x}_{k-1})\, dt , \end{aligned} \tag{4.2.112}$$

with

$$\nabla f(\underline{x}_{k-1}) + \hat{G}_{k-1}(\underline{x}_k - \underline{x}_{k-1}) = \nabla\hat{q}_{k-1}(\underline{x}_k) .$$

Since $\tau_k = \alpha_k = 1$, $\underline{x}_p = \underline{x}_{k+1}$ and with $\alpha_k = 1$ the inequality

$$<\nabla\hat{q}_{k-1}(\underline{x}_k), \underline{x}_{k+1} - \underline{x}_k> \geq 0 , \quad \underline{x}_{k+1} \in R$$

follows from the optimality of $\underline{x}_k$ for $\hat{q}_{k-1}$. Taking the scalar product of (4.2.112) with $\underline{x}_{k+1} - \underline{x}_k$ and using (4.2.111), yields

$$- \hat{q}_k(\underline{x}_{k+1}) \leq - \int_0^1 <\underline{x}_{k+1} - \underline{x}_k, \{G[\underline{x}_{k-1} + t(\underline{x}_k - \underline{x}_{k-1})] - G_c + G_c - \hat{G}_{k-1}\} (\underline{x}_k - \underline{x}_{k-1})> dt$$

$$\leq (\tfrac{1}{2} ||G_c - \hat{G}_{k-1}|| + \mu^c_{k-1}) ||\underline{x}_{k+1} - \underline{x}_k|| \, ||\underline{x}_k - \underline{x}_{k-1}|| . \quad (4.2.113)$$

Since $\frac{1}{2} ||G_c - \hat{G}_k|| + \mu^c_k < \ell/2$, and using (4.2.83) with $\alpha_k = \tau_k = 1$ yields

$$||\underline{x}_{k+1} - \underline{x}_k|| \leq r \, ||\underline{x}_k - \underline{x}_{k-1}|| \quad (4.2.114)$$

where $r \in (0,1)$ and is given by

$$r = \frac{1}{\ell} (\tfrac{1}{2} ||G_c - \hat{G}_{k-1}|| + \mu^c_{k-1}) < 1 .$$

Proceeding as in (4.2.67), the inequality

$$||\underline{x}_c - \underline{x}_k|| \leq \frac{r}{1-r} ||\underline{x}_k - \underline{x}_{k-1}||$$

is obtained.

If for $k > K_o$ the conditions (4.2.105) and $||\underline{x}_k - \underline{x}_c|| < t_1$ are satisfied, the above inequality becomes

$$||\underline{x}_k - \underline{x}_c|| \leq \frac{r^{k-k_o}}{1-r} ||\underline{x}_{K_o+1} - \underline{x}_{K_o}|| \quad (4.2.115)$$

which, by definition establishes the R-linear convergence of $\{\underline{x}_k\}$

(see Ortega and Rheinboldt (1970)). The scalar t_1 is chosen to satisfy $\frac{1}{1-r} \|\underline{x}_{K_0+1} - \underline{x}_{K_0}\| \leq t_1$ and $\|\underline{x}_{K_0} - \underline{x}_c\| \leq t_1$ such that $\mu^c_{K_0}$ is small enough to satisfy (4.2.110) for given $\hat{G}_{K_0}$. The former inequality and (4.2.113) - (4.2.115) ensure that

$$\|\underline{x}_c - \underline{x}_{K_0+1}\| \leq \frac{1}{1-r} \|\underline{x}_{K_0+1} - \underline{x}_{K_0}\| < t_1 .$$

Thus, by induction, if for $K_0 \leq j \leq k-1$, $\|\underline{x}_j - \underline{x}_c\| \leq t_1$ holds, $\|\underline{x}_k - \underline{x}_c\| < t_1$ also holds. Hence $\|\underline{x}_{K_0} - \underline{x}_c\| \leq t_1$ ensures $\|\underline{x}_k - \underline{x}_c\| < t_1$, $\forall\, k > K_0$. Therefore, R-linear convergence for $k \geq K_0$ follows from (4.2.115) if the conditions (4.2.105) are satisfied.

The linear convergence of $\{\underline{x}_k\}$, generated by (4.2.101), (4.2.102) and the stepsize strategy (4.2.104), may be shown using (4.2.48). Thus

$$f(\underline{x}_{k+1}) - f(\underline{x}_k) \leq \tau_k \langle \nabla f(\underline{x}_k), \underline{x}_p - \underline{x}_k \rangle \left(1 - \frac{\tau_k \alpha_k}{2} - \frac{\tau_k \alpha_k}{\ell} \left(\tfrac{1}{2}\|G_c - \hat{G}_k\| + \mu^c_k\right)\right)$$

and for α_k set to unity for $k \geq K_0$

$$\tfrac{1}{2}\|G_c - \hat{G}_k\| + \mu^c_k < \ell/2$$

is satisfied and the inequalities

$$0 \leq c_1 \leq 1 - \frac{\tau_k}{2} - \frac{\tau_k}{\ell}\left(\tfrac{1}{2}\|G_c - \hat{G}_k\| + \mu^c_k\right) < 1 - \frac{\bar{\alpha}}{2} \leq \tfrac{1}{2}$$

hold for $\tau_k = 1$. The inequality

$$\langle \nabla f(\underline{x}_k), \underline{x}_{k+1} - \underline{x}_k \rangle \geq -\left(\tfrac{1}{2}\|G_c - \hat{G}_k\| + \mu^c_k\right) \|\underline{x}_{k+1} - \underline{x}_k\| \, \|\underline{x}_k - \underline{x}_{k-1}\|$$

follows from (4.2.112) for $\tau_k = 1$ and $\underline{x}_k = \underline{x}_p$. Using (4.2.111), the rest of the linear convergence rate proof is the same as that of stepsize strategy (4.2.103).

To establish the superlinear convergence of $\{x_k\} \subset S$, given $\alpha_k = \tau_k = 1$, consider

$$\underline{x}_p - \underline{x}_k = \underline{x}_{k+1} - \underline{x}_k = \bar{P}_k(\underline{x}_{k+1} - \underline{x}_k) \; . \tag{4.2.116}$$

The second order expansion of $f(\underline{x})$ at $\underline{x}_k$ may be written as

$$f(\underline{x}_k) = f(\underline{x}_{k-1}) + \hat{q}_{k-1}(\underline{x}_k) + \int_0^1 (1-t) < \bar{P}_{k-1}(\underline{x}_k - \underline{x}_{k-1}), \{G[\underline{x}_{k-1} + t(\underline{x}_k - \underline{x}_{k-1})] - G_c + G_c - \hat{G}_k\} \bar{P}_{k-1}(\underline{x}_k - \underline{x}_{k-1}) > dt \tag{4.2.117}$$

thus, (4.2.112) may be rewritten as

$$\nabla f(\underline{x}_k) = \nabla \hat{q}_{k-1}(\underline{x}_k) + \int_0^1 \{G[\underline{x}_{k-1} + t(\underline{x}_k - \underline{x}_{k-1})] - G_c + G_c - \hat{G}_k\} \bar{P}_{k-1}(\underline{x}_k - \underline{x}_{k-1}) dt \tag{4.2.118}$$

hence, using $P_k(\underline{x}_{k+1} - \underline{x}_k) = \underline{x}_{k+1} - \underline{x}_k$, (4.2.113) becomes

$$- \hat{q}_k(\underline{x}_{k+1}) \le \left(\frac{\| \bar{P}_k \{G_c - \hat{G}_{k-1}\} \bar{P}_{k-1}(\underline{x}_k - \underline{x}_{k-1}) \|}{\| \underline{x}_k - \underline{x}_{k-1} \|} + \int_0^1 \| \bar{P}_k \{G[\underline{x}_{k-1} + t(\underline{x}_k - \underline{x}_{k-1})] - G_c\} \bar{P}_{k-1} \| dt \right) \| \underline{x}_k - \underline{x}_{k-1} \| \| \underline{x}_{k+1} - \underline{x}_k \| \tag{4.2.119}$$

and using (4.2.83) with $\alpha_k = 1$ yields

$$\| \underline{x}_{k+1} - \underline{x}_k \| \le \frac{2}{\ell} (\cdot) \; \| \underline{x}_k - \underline{x}_{k-1} \| \tag{4.2.120}$$

where $(\cdot)$ is the term in parenthesis in (4.2.119). Since $\mu_k^c \to 0$ the second term in the parenthesis approaches zero in the limit. Furthermore, by hypothesis, (4.2.106) ensures that, in the limit, the first term in the parenthesis also becomes zero. Q-superlinear convergence follows identically as in (4.2.67) - (4.2.69) of Theorem (4.8). This result holds for (4.2.104) as well as (4.2.103). For (4.2.104) the above linear convergence argument can be extended, as in Theorem (4.6), to superlinear convergence, for (.) given in (4.2.120), converging to zero in the limit. This completes the proof of Theorem (4.9). □

Clearly, steps (4.2.101) - (4.2.103) or (4.2.101),(4.2.102), (4.2.104) are different implementations of algorithm (4.1.7).

Condition (4.2.106) requires the projection of $\hat{G}_k$ onto $N^T(\underline{x}_{k+1},\underline{x}_k)(\underline{x}_{k+1} - \underline{x}_k) = \underline{o}$ to be close to the projection of G_c. This is a weaker requirement on $\hat{G}_k$ than imposing $\hat{G}_k$ itself to be close to G_c. An R-linear convergence proof for algorithm (4.1.7) can also be given by viewing this algorithm as an unconstrained optimisation algorithm for a lower dimensional problem over the constraints active at $\underline{x}_c$. If this approach is taken, a result such as that given by Ortega and Rheinboldt (1970, Theorem 14.1.6) can be used to show R-linear rate of convergence without assuming that $\hat{G}_k$ is chosen to be near G_c. However, the above constructive approach for deriving condition (4.2.106) was preferred.

A condition similar to (4.1.106), involving the Hessian of the Lagrangian (1.2.17) with respect to $\underline{x}$, is discussed by Powell(1977,b) in connection with the nonlinearly constrained optimisation algorithms discussed in Powell(1977,a) and Han(1977).

In general, a positive definite matrix cannot be chosen to be arbitrarily close, in a given norm, to a fixed matrix. Thus, in general, (4.2.105,b) need not be satisfied unless G_c is positive definite. If G_c is assumed to be positive definite, then (4.2.106) may also be written as

$$\lim_{k\to\infty} \frac{\|(\hat{G}_k - G_c)(\underline{x}_{k+1} - \underline{x}_k)\|}{\|\underline{x}_{k+1} - \underline{x}_k\|} = 0 \qquad (4.2.121)$$

which is the superlinear convergence condition originally developed for unconstrained optimisation algorithms using quasi-Newton updating schemes for $\hat{G}_k$ (Dennis and Moré (1974)). The derivation of (4.2.121) within Theorem (4.9) can be done by using (4.2.116) and replacing $\bar{P}_{k-1}(\underline{x}_k - \underline{x}_{k-1})$ in (4.2.117) - (4.2.119) with $\underline{x}_k - \underline{x}_{k-1}$. Condition (4.2.121) contrasts with (4.2.106) in which only projections of $\hat{G}_k$ and G_c are measured. The projection of $\hat{G}_k$ may become arbitrarily close to the projection of G_c even when $\hat{G}_k$ is positive definite and G_c is not. It should be noted that if G_c is positive definite, the assumption $\{\underline{x}_k\} \subset S$ need not be made since $\bar{P}_{k-1}$ is no longer needed in (4.2.116)-(4.2.119). Furthermore, in linearly constrained optimisation $\bar{P}_k$ is fixed for $\{\underline{x}_k\} \subset S$ since the active set remains unaltered until $\underline{x}_c$ is reached.

If for some $K_o \geq o$, $\hat{G}_{K_o}$ satisfies (4.2.105,b), then the following Lemma gives a sufficient condition on the updates of $\hat{G}_k$ which ensures that the bound (4.2.105,b) is satisfied for all $k \geq K_o$. This condition has been studied for unconstrained optimisation problems by Broyden, Dennis and More (1973). Let $\|.\|'$ denote any fixed matrix norm which may be different from $\|.\|$ which is the operator norm induced by the corresponding vector norm.

Lemma (4.9)

Let $f(\underline{x})$ have a Lipschitz continuous second derivative at $\underline{x}_c$. Let $\underline{x}_c$ be the optimum of $f(\underline{x})$ over the feasible region R . If there are two nonnegative numbers α_1 and α_2 such that

$$\|\hat{G}_{k+1} - G_c\|' \le (1+\alpha_1 \max\{\|\underline{x}_{k+1} - \underline{x}_c\|, \|\underline{x}_k - \underline{x}_c\|\})\|\hat{G}_k - G_c\|' + \alpha_2 \max\{\|\underline{x}_{k+1} - \underline{x}_c\|, \|\underline{x}_k - \underline{x}_c\|\} \quad (4.2.122)$$

then for any $r \in (0,1)$ and $t_2 > o$ there exist two positive numbers $\bar{t}_1(r,t_2)$ and $\bar{t}_2(r,t_2)$ such that for $K_o \ge o$ if $\|\underline{x}_{K_o} - \underline{x}_c\| \le \bar{t}_1(r,t_2)$ and $\|\hat{G}_{K_o} - G_c\| \le \bar{t}_2(r,t_2)$ then the sequence of matrices $\hat{G}_k$ satisfy $\|\hat{G}_k - G_c\| \le t_2$ for each $k \ge K_o$ and $\{\underline{x}_k\}$, generated by (4.2.101)-(4.2.103) or (4.2.101),(4.2.102),(4.2.104) converges R-linearly to $\underline{x}_c$.

Proof

Since all norms in a finite dimensional vector space are equivalent, there exist two positive numbers d and d' such that for any $n \times n$ matrix A

$$d\|A\|' \ge \|A\| , \quad d'\|A\| \ge \|A\|' . \quad (4.2.123)$$

Let $r \in (0,1)$ be given. By Theorem (4.9), there exist two positive numbers t_1 and t_2 such that if for $k \ge K_o \ge o$, $\|\underline{x}_{K_o} - \underline{x}_c\| \le t_1$ and if $\hat{G}_k$ satisfies $\|\hat{G}_k - G_c\| \le t_2$ then $\|\underline{x}_k - \underline{x}_c\| \le \varepsilon r^{k-K_o}$, $\varepsilon > o$, $\forall k$ satisfying these conditions.

Let two positive numbers $\bar{t}_1 = \bar{t}_1(r,t_2)$, $\bar{t}_2 = \bar{t}_2(r,t_2)$ be chosen such that the following inequalities are satisfied:

$$\bar{t}_1 \le t_1 \qquad (4.2.124,a)$$

$$2\,d\,d'\,\bar{t}_2 \le t_2 \qquad (4.2.124,b)$$

$$(2\,\alpha_1\,\bar{t}_2\,d' + \alpha_2)\,\frac{\bar{t}_1}{1-r} \le d'\,\bar{t}_2\,. \qquad (4.2.124,c)$$

Let $\|\underline{x}_{K_0} - \underline{x}_c\| \le \bar{t}_1$ and $\|\hat{G}_{K_0} - G_c\| \le \bar{t}_2$. If it can be shown that for $j \ge K_0$

$$\|\underline{x}_j - \underline{x}_c\| \le r^{j-K_0}\,\bar{t}_1 \qquad (4.2.125,a)$$

$$\|\hat{G}_j - G_c\|' \le 2\,d'\,\bar{t}_2 \qquad (4.2.125,b)$$

then by (4.2.124,a) and (4.2.124,b), $\|\underline{x}_j - \underline{x}_c\| \le t_1$ and $\|\hat{G}_j - G_c\| \le 2\,d\,d'\,\bar{t}_2 \le t_2$ thus, this lemma follows immediately from Theorem (4.9). Inequalities (4.2.125) will be established by induction. Clearly (4.2.125) hold for $j = o$. Assume that they are true for j, $o \le K_0 \le j \le k$. By the induction hypothesis and using (4.2.124,a),(4.2.124,b), the inequalities $\|\underline{x}_k - \underline{x}_c\| \le r^{k-K_0}\bar{t}_1 \le t_1$, $k \ge K_0$, and $\|\hat{G}_k - G_c\| \le 2\,d\,d'\bar{t}_2 \le t_2$ follow. Hence it follows from Theorem (4.11) that for $k \ge K_0 \ge o$, $\|\underline{x}_{k+1} - \underline{x}_c\| \le r^{k-K_0+1}\,\bar{t}_1$. Therefore (4.2.125,a) is true for $j = k+1$. It also follows from the induction hypothesis and (4.2.122) that (assuming for simplicity $K_0 = o$)

$$\|\hat{G}_{j+1} - G_c\|' - \|\hat{G}_j - G_c\|' \le 2\alpha_1 d'\,\bar{t}_2\bar{t}_1 r^j + \alpha_2\bar{t}_1 r^j\,.$$

Summing this expression from $j=o$ to $j=k$ yields

$$\|\hat{G}_{k+1} - G_c\|' \le \|\hat{G}_o - G_c\|' + (2\alpha_1 d'\bar{t}_2 + \alpha_2)\,\frac{\bar{t}_1}{1-r}\,.$$

By (4.2.124,c) and (4.2.123) and the initial choice of $\hat{G}_0$ this can be written as

$$\|\hat{G}_{k+1} - G_c\|' \le d'\bar{t}_2 + d'\bar{t}_2 = 2\,d'\bar{t}_2$$

$$\|\hat{G}_{k+1} - G_c\| \le 2\,d\,d'\bar{t}_2 \le t_2 \qquad (4.2.126)$$

Hence (4.2.125,b) is also established. Since Theorem (4.11) establishes the R-linear convergence of $\{\underline{x}_k\}$ if $\|\hat{G}_k - G_c\| \le t_2$, $\|\underline{x}_{K_0} - \underline{x}_c\| \le t_1$, by choosing $\bar{t}_1, \bar{t}_2$ to satisfy $\bar{t}_1 \le t_1$, $2dd'\bar{t}_2 \le t_2$ the required result is established. □

To show that condition (4.2.122) is satisfied for some updates, the following results are necessary:

<u>Lemma (4.10)</u> (Dennis and More (1974, Lemma 3.3)

Let $\{a_k\}$ and $\{b_k\}$ be sequences of nonnegative numbers such that

$$a_{k+1} \le (1 + \alpha_1 b_k) a_k + \alpha_2 b_k$$

for $\alpha_1 \ge 0$, $\alpha_2 \ge 0$ and $\sum_{k=1}^{\infty} b_k < \infty$, then $\{a_k\}$ converges.

□

<u>Lemma (4.11)</u>

Let $\underline{x}_c$ be the constrained minimum of f over R and let G_c be positive definite. Assume that f and $\{\underline{x}_k\}$ satisfy the hypotheses of Theorem (4.9). If there exist two nonnegative constants α_1 and α_2 and two sequences of nonnegative numbers $\{\rho_k\}$ and $\{\sigma_k\}$ such that the following conditions are satisfied

(i) $\sum_{k=1}^{\infty} \|\underline{x}_k - \underline{x}_c\| < \infty$;

(ii) $\rho_k \to 0$ or $\sigma_k \to 0$ implies $\dfrac{\|(\hat{G}_k - G_c)(\underline{x}_{k+1} - \underline{x}_k)\|}{\|\underline{x}_{k+1} - \underline{x}_k\|} \to 0$

(iii) $\rho_{k+1} \le (1 - \sigma_k + \alpha_1 \max\{\|\underline{x}_{k+1} - \underline{x}_c\|, \|\underline{x}_k - \underline{x}_c\|\})\, \rho_k$

$\qquad + \alpha_2 \max\{\|\underline{x}_{k+1} - \underline{x}_c\|, \|\underline{x}_k - \underline{x}_c\|\}$

then $\{\underline{x}_k\}$ converges Q-superlinearly to $\underline{x}_c$.

Proof

The proof is analogous to a similar result in Han (1976, Theorem 4.5). From (i) and (iii) and Lemma (4.10) it follows that $\{\rho_k\}$ converges to a nonnegative number, say $\bar{\rho}$. If $\bar{\rho}$ is zero then by (ii), (4.2.121) and Theorem (4.11) the desired conclusion follows. Assuming $\bar{\rho} \neq o$, (iii) implies that

$$\sigma_k \rho_k \leq \rho_k - \rho_{k+1} + \max\{ \|\underline{x}_{k+1} - \underline{x}_c\|, \|\underline{x}_k - \underline{x}_c\| \} (\alpha_1 \rho_k + \alpha_2) . \tag{4.2.127}$$

By taking limits of both sides of (4.2.127) and using (i) and $\sigma_k \rho_k \geq o$, it follows that $\sigma_k \rho_k \to o$. Since $\rho_k \to \bar{\rho} \neq o$, it follows that $\sigma_k \to o$. Hence by (ii) and Theorem (4.9) the desired conclusion follows. □

Clearly, (ii) above could also have been written in terms of condition (4.2.106) of Theorem (4.9) if the positive definiteness of G_c is relaxed but $\hat{G}_k$ is restricted to be positive definite. If neither matrix is restricted to be positive definite, the (ii) is adequate. Another justification for the formulation in (ii) is that, if $\hat{G}_k$ is restricted to be positive definite, (4.2.105,b) is in general true for positive definite G_c .

Consider the family of updates (see Dennis (1972))

$$\hat{G}_{k+1} = \hat{G}_k + \frac{(\underline{\gamma} - \hat{G}_k\underline{\delta})\underline{c}^T + \underline{c}(\underline{\gamma} - \hat{G}_k\underline{\delta})^T}{\langle \underline{c}, \underline{\delta} \rangle} - \frac{\underline{\delta}^T(\underline{\gamma} - \hat{G}_k\underline{\delta})\underline{c}\,\underline{c}^T}{\langle \underline{c}, \underline{\delta} \rangle^2} \tag{4.2.128}$$

where $\underline{\gamma},\underline{\delta}$ are given by (4.2.23,b,c) and $\underline{c}$ is any vector such that $<\underline{c},\underline{\delta}> \neq 0$. Clearly, each choice of $\underline{c}$ determines a particular update. For unconstrained optimisation, the convergence of these updates have been studied by Broyden, Dennis and Moré (1973), Dennis and Moré (1974). For constrained optimisation, Han (1976) has studied the convergence of $\hat{G}_k$ as the approximation to the Hessian of the Lagrangian (1.2.17). Three choices of $\underline{c}$ are studied below. The first is $\underline{c} = D\,\underline{\delta}$ where D is any fixed positive definite matrix. The second is $\underline{c} = \underline{\delta}$ which can also be obtained by setting $D = I$ in the first choice. In the unconstrained case $\underline{c} = \underline{\delta}$ reduces (4.2.128) to Powell's (1970,a,b) update. The third choice $\underline{c} = \underline{\gamma}$ yields the DFP update. The DFP update for $\hat{H}_k$ was discussed in Section (4.1.2).

To show that (4.2.128) satisfies the requirements for superlinear convergence stated in Theorem (4.11), Lemmas (4.9),(4.11), the following Lemma is introduced.

Lemma (4.12) (Broyden, Dennis and Moré (1973, Lemma 5.2), Han (1976, Lemma 5.1))

Let $\hat{G}_k$ be any $n \times n$ symmetric matrix and $\underline{\delta}, \underline{c}, \underline{\gamma} \in E^n$ with $<\underline{c}, \underline{\delta}> \neq 0$ and define $\hat{G}_{k+1}$ by (4.2.128). Let M be a $n \times n$ nonsingular, symmetric matrix and let $\|.\|_M$ be the matrix norm defined by

$$\|Q\|_M = \text{trace } [(MQM)^T (MQM)] . \qquad (4.2.129)$$

If M satisfies

$$\|M\,\underline{c} - M^{-1}\,\underline{\delta}\| \leq \beta\,\|M^{-1}\,\underline{\delta}\| \qquad (4.2.130)$$

for some $\beta \in [0,\frac{1}{3}]$ then for any symmetric $n \times n$ matrix G_c we have

$$\|\hat{G}_{k+1} - G_c\|_M \le ((1-\lambda\theta^2)^{\frac{1}{2}} + \lambda_1 \frac{\|M\,\underline{c} - M^{-1}\,\underline{\delta}\|}{\|M^{-1}\underline{\delta}\|})\,\|\hat{G}_k - G_c\|_M + \lambda_2 \frac{\|\underline{\gamma} - G_c\,\underline{\delta}\|}{\|\underline{\delta}\|} \qquad (4.2.131)$$

where $\lambda \in (0,1]$ and λ_1, λ_2 are constants that depend on M and n and

$$\theta = \frac{\|M(\hat{G}_k - G_c)\,\underline{\delta}\|}{\|\hat{G}_k - G_c\|_M\,\|M^{-1}\underline{\delta}\|} \le 1 \qquad (4.2.132)$$

if $\hat{G}_k \ne G_c$ and $\theta = o$ otherwise. □

Theorem (4.10) below establishes a sufficient condition for Q-superlinear convergence. The proof is based on the superlinear convergence results of Broyden, Dennis and Moré (1973), Dennis and Moré (1974) and Han (1976).

The choice $\underline{c} = \underline{\gamma}$ in (4.2.128), yielding the DFP formula, is one case in which $\hat{G}_k$ is ensured to remain positive definite provided $\langle\underline{\delta},\underline{\gamma}\rangle > o$. However, $\hat{G}_k$ computed using (4.2.128) is not, in general, ensured to remain positive definite. This is so, for example, when $\underline{c} = \underline{\delta}$. Whenever necessary, the positive definitness of Hessian approximations may be ensured by using

$$\hat{G}_{k+1} = \hat{G}(\Xi) = \hat{G}_k + \frac{\Xi[(\underline{\gamma} - \hat{G}_k\,\underline{\delta})\underline{c}^T + \underline{c}(\underline{\gamma} - \hat{G}_k\underline{\delta})^T]}{\langle\underline{c},\underline{d}\rangle} - \frac{\Xi^2[\underline{\delta}^T(\underline{\gamma} - \hat{G}_k\,\underline{\delta})\,\underline{c}\,\underline{c}^T]}{\langle\underline{c},\underline{\delta}\rangle^2}$$

with the scalar Ξ chosen such that $\hat{G}_{k+1}$ remains positive definite. For $\underline{c} = \underline{\delta}$, Powell (1970,a) suggests the use of the bound

$|\Xi - 1| \leq \hat{\Xi} < 1$ in setting Ξ to ensure $|\det \hat{G}_{k+1}| \geq .1|\det \hat{G}_k|$ in unconstrained optimisation problems such that $\hat{G}_{k+1}$ is nonsingular. Powell (1970,b) describes how this can be done. Moré and Trangenstein (1976) prove the superlinear convergence of unconstrained optimisation methods for $\underline{c} = \underline{\delta}$, with Ξ such that $\hat{G}_{k+1}$ is nonsingular and $|\Xi - 1| \leq \hat{\Xi} < 1$. For linearly constrained optimisation problems Fletcher (1972) suggests the use of the above formula with $\underline{c} = \underline{\delta}$ and $\Xi \in [0,1]$ chosen to ensure that $\det \hat{G}_{k+1}$ is not less than $.1 \det \hat{G}_k$. Keeping $\det \hat{G}_{k+1}$ positive ensures the positive definiteness of $\hat{G}_{k+1}$ since it can be shown that not more than one eigenvalue can be introduced by applying the above updating formula with $0 \leq \Xi \leq 1$ (see, Wilkinson (1965, pp. 97 - 98)).

The convergence results in this section are not altered if (4.2.128) is replaced by the above modification. This is because Lemma (4.12) is also valid for the above formula. Only the values of the constants $\lambda, \lambda_1, \lambda_2$ in (4.2.131) are affected by Ξ such that for $\beta \in [0, {}^1/3]$ and for a general choice of $\underline{c}$ ($\langle \underline{c}, \underline{\delta} \rangle > 0$),

$$\lambda = \frac{(2 - \Xi - 2\beta)\Xi}{1 - \beta^2}$$

$$\lambda_1 = \frac{5}{2} \Xi (1 - \beta)^{-1}$$

$$\lambda_2 = 2 \Xi (1 + \frac{3}{2} \sqrt{n}) (\text{trace}[M^T M])$$

These constants have been derived directly by replacing (4.2.128) with its modification above, in results due to Broyden, Dennis and Moré (1973, Lemmas 5.1 - 5.2). For $\Xi \in [0,1]$, $\lambda \in [0,1]$, for $\Xi \in (0,1]$, $\lambda \in (0,1]$ and for $\Xi \in (0,2)$ implied by $|\Xi - 1| \leq \hat{\Xi} < 1$, the bound $\lambda \le 1$ is ensured. The superlinear convergence results in this section are not altered provided $\lambda \le 1$ and hence $\Xi \in (0,2)$. With a general choice for $\underline{c}$ ($\langle \underline{c}, \underline{\delta} \rangle > o$), $\hat{G}_{k+1}$ can be ensured to remain positive definite if $\Xi = \Xi_1$ and Ξ_1 is chosen such that a condition like $\det \hat{G}_{k+1} = \det \hat{G}(\Xi_1) \ge .1 \det \hat{G}_k$ is satisfied and $\det \hat{G}(\Xi) \ne o \;\forall\, \Xi \in [0, \Xi_1]$, hence $\hat{G}_{k+1}$ is nonsingular. This follows from a theorem due to Caratheodory (1967, Theorem 6, p.190) which states that a matrix $\hat{G}(\Xi)$ is positive definite over $\Xi \in [\Xi_\ell, \Xi_u]$ if $\hat{G}(\Xi)$ is nonsingular over this interval and for $\Xi_o \in [\Xi_\ell, \Xi_u]$, $\hat{G}(\Xi_o)$ is positive definite. Setting $\Xi_\ell = \Xi_o = o$, $\Xi_u = \Xi_1$ yields above the condition.

Theorem (4.10)

Let $\underline{x}_c$ be the minimum of f over R, let f and R satisfy the hypotheses of Theorem (4.9) and let f have a Lipschitz continuous second derivative at $\underline{x}_c$. Suppose that the sequence of matrices $\{\hat{G}_k\}$ are generated by the update (4.2.128) with any $\underline{c}$ such that $\underline{c}^T \underline{\delta} \ne o$,

$$\frac{\|M \underline{c} - M^{-1} \underline{\delta}\|}{\|M^{-1} \underline{\delta}\|} \le \mu \max \{\|\underline{x}_k - \underline{x}_c\|, \|\underline{x}_{k+1} - \underline{x}_c\|\} \tag{4.2.133}$$

for a constant μ and an arbitrary but fixed nonsingular symmetric matrix M. If for $K_o \ge o$ $\underline{x}_{K_o}$ and $\hat{G}_{K_o}$ are sufficiently close to $\underline{x}_c$, G_c respectively, then the sequence $\{\underline{x}_k\}$, generated by the algorithm described in (4.2.101) - (4.2.104), is well defined and converges Q-superlinearly to $\underline{x}_c$.

Proof

The proof is similar to a similar result due to Han (1976, Theorem 5.2) for the convergence of approximations to the Hessian of the Lagrangian (1.2.17).

For any $r \in (0,1)$ and any fixed $t_2 > o$, let $\bar{t}_1 = \bar{t}_1\ (r,t_2)$, $\bar{t}_2 = \bar{t}_2\ (r,t_2)$ be defined as in Lemma (4.9) with the matrix norm $\|.\|'$ as $\|.\|_M$, and let

$$\alpha_1 = \lambda_1 \mu \qquad \alpha_2 = \lambda_2 K \tag{4.2.134}$$

where λ_1 and λ_2 are defined in Lemma (4.12) and K is the Lipschitz constant for the second derivative of f at $\underline{x}_c$. Furthermore, $\bar{t}_1$ is required to satisfy

$$\bar{t}_1 \leq \frac{1}{3\mu} \ . \tag{4.2.135}$$

First, it will be demonstrated by induction that if $\|\underline{x}_{K_0} - \underline{x}_c\| \leq \bar{t}_1$ and $\|\hat{G}_{K_0} - G_c\| \leq \bar{t}_2$ then the generated sequence $\{\underline{x}_j\}$ exists and converges to $\underline{x}_c$ at least R linearly; that is

$$\|\underline{x}_j - \underline{x}_c\| \leq \bar{t}_1\ r^{j-K_0} \qquad , \ \forall_j \geq K_0 \geq o\ . \tag{4.2.136}$$

When $j = o$, the existence of $\underline{x}_{j+1}$ satisfying (4.2.136) follows from the choice $\bar{t}_1$ and $\bar{t}_2$ and Theorem (4.9). Assume that for all $j \leq k$, $k \geq K_0$, $\underline{x}_{j+1}$ exists and (4.2.136) holds; it is shown that $\underline{x}_{k+2}$ exists and (4.2.136) is also true for j=j+1.

If $\underline{\delta} = \underline{o}$ then $\underline{x}_{k+1} = \underline{x}_c$ (which is a unique minimum in $\{\underline{x} \in E^n \mid \|\underline{x} - \underline{x}_c\| < \bar{t}_1\}$ if R is convex) and the sequence $\{\underline{x}_k\}$ converges to $\underline{x}_c$ in a finite number of steps. When $\underline{\delta} \neq o$

it follows from (4.2.133) and (4.2.135) that

$$\frac{\|M\,\underline{c} - M^{-1}\,\underline{\delta}\|}{\|M^{-1}\,\underline{\delta}\|} \le \mu \max\{\|\underline{x}_k - \underline{x}_c\|, \|\underline{x}_{k+1} - \underline{x}_c\|\} < \mu\,\bar{t}_1 \le \frac{1}{3} . \tag{4.2.137}$$

Hence by (4.2.131) of Lemma (4.12) the following inequality is obtained

$$\|\hat{G}_{k+1} - G_c\|_M \le ((1 - \lambda\,\theta^2)^{\frac{1}{2}} + \lambda_1\mu \max\{\|\underline{x}_k - \underline{x}_c\|, \|\underline{x}_{k+1} - \underline{x}_c\|\}) \\ \|\hat{G}_k - G_c\|_M + \frac{\lambda_2\,\|\,\underline{\gamma} - G_c\,\underline{\delta}\,\|}{\|\underline{\delta}\|} \tag{4.2.138}$$

where θ is given by (4.2.132).

The inequality

$$\|\nabla f(\underline{x}_{k+1}) - \nabla f(\underline{x}_k) - G_c(\underline{x}_{k+1} - \underline{x}_k)\| \\ \le \sup_{0\le s\le 1} \|G[\underline{x}_k + s(\underline{x}_{k+1} - \underline{x}_k)] - G_c\|\,\|\underline{x}_{k+1} - \underline{x}_k\|$$

holds as an application of the mean value theorem (Ortega and Rheinboldt (1970, Theorem 3.2.5)), in a neighbourhood of $\underline{x}_c$, i.e. $\forall\,\underline{x} \in \{\underline{x} \in E^n \mid \|\underline{x} - \underline{x}_c\| \le \bar{t}_1\}$. The Lipschitz continuity of the second derivative at $\underline{x}_c$, with constant K, may be used to express this inequality as

$$\|\underline{\gamma} - G_c\,\underline{\delta}\| \le K\|\underline{\delta}\| \sup_{0\le s\le 1} \|\underline{x}_k + s(\underline{x}_{k+1} - \underline{x}_k) - \underline{x}_c\| \\ = K\|\underline{\delta}\| \max\{\|\underline{x}_k - \underline{x}_c\|, \|\underline{x}_{k+1} - \underline{x}_c\|\}$$

and therefore

$$\frac{\|\underline{\gamma} - G_c\,\underline{\delta}\|}{\|\underline{\delta}\|} \le K \max\{\|\underline{x}_{k+1} - \underline{x}_c\|, \|\underline{x}_k - \underline{x}_c\|\} . \tag{4.2.139}$$

By (4.2.138), (4.2.139) and using (4.2.134) the following inequality is obtained

$$\|\hat{G}_{k+1} - G_c\|_M \leq ((1 - \lambda\,\theta^2)^{\frac{1}{2}} + \alpha_1 \max\{\|\underline{x}_{k+1} - \underline{x}_c\|, \|\underline{x}_k - \underline{x}_c\|\}) \|\hat{G}_k - G_c\|_M + \alpha_2 \max\{\|\underline{x}_{k+1} - \underline{x}_c\|, \|\underline{x}_k - \underline{x}_c\|\} \; . \tag{4.2.140}$$

The existence of $\underline{x}_{k+2}$ satisfying $\|\underline{x}_{k+2} - \underline{x}_c\| \leq \bar{t}_1 r^{k+2-K_0}$, $k \geq K_0 \geq o$, follows from (4.2.140) and Lemma (4.9) immediately. This demonstrates the R-linear convergence of $\{\underline{x}_k\}$ to $\underline{x}_c$. To show that $\{\underline{x}_k\}$ converges at a Q-superlinear rate, Lemma (4.11) will be used. Let $\rho_k = \|\hat{G}_k - G_c\|_M$ and $\sigma_k = \frac{1}{2}\lambda\,\theta^2$ where θ is defined for each k , by (4.2.132). Condition (i) of Lemma(4.11) is satisfied because $\{\underline{x}_k\}$ converges R-linearly to $\underline{x}_c$. Clearly, $\rho_k \to o$ implies

$$\lim_{k\to\infty} \frac{\|(\hat{G}_k - G_c)\,\underline{\delta}\|}{\|\underline{\delta}\|} = o \tag{4.2.141}$$

and so does $\sigma_k \to o$. Hence by (4.2.141) condition (ii) of Lemma (4.11) is satisfied. By taking the inequality $(1 - \lambda\,\theta^2)^{\frac{1}{2}} \leq 1 - \frac{1}{2}\lambda\,\theta^2$ into account, it follows from (4.2.140) that

$$\rho_{k+1} \leq (1 - \sigma_k + \alpha_1 \max\{\|\underline{x}_k - \underline{x}_c\|, \|\underline{x}_{k+1} - \underline{x}_c\|\})\,\rho_k + \alpha_2 \max\{\|\underline{x}_k - \underline{x}_c\|, \|\underline{x}_{k+1} - \underline{x}_c\|\} \; .$$

Thus all conditions of Lemma (4.11) are satisfied and hence $\{\underline{x}_k\}$ converges Q-superlinearly to $\underline{x}_c$. □

Clearly, an update of the type (4.2.128) is required to satisfy the condition (4.2.133) in order to ensure superlinear convergence. When $\underline{c}$ is given by $\underline{c} = D\,\underline{\delta}$, for any fixed positive definite matrix D, then choosing $M = (D)^{-\frac{1}{2}}$ satisfies (4.2.130) and ensures that (4.2.133) is satisfied. Since $\underline{c} = \underline{\delta}$ for Powell's (1970 a,b) update implies $D = I$, (4.2.133) is satisfied. For the DFP update, with $\underline{c} = \underline{\gamma}$, assuming G_c to be positive definite, choosing $M = (G_c)^{-\frac{1}{2}}$ and using the mean value theorem (Ortega and Rheinboldt (1970, Theorem 3.2.5)) yields

$$\|M\underline{c} - M^{-1}\underline{\delta}\| \le \|M\|\ \|\underline{\gamma} - G_c\underline{\delta}\|$$

$$\le K\|M\| \max\{\|\underline{x}_k - \underline{x}_c\|, \|\underline{x}_{k+1} - \underline{x}_c\|\}\ \|\underline{\delta}\| .$$

Hence

$$\frac{\|M\underline{c} - M^{-1}\underline{\delta}\|}{\|M^{-1}\underline{\delta}\|} \le K\|M\|^2 \max\{\|\underline{x}_k - \underline{x}_c\|, \|\underline{x}_k - \underline{x}_c\|\}$$

and (4.2.133) is satisfied with $\mu = K\|M\|^2$ where K is the Lipschitz constant for the Hessian of f at $\underline{x}_c$. □

In constrained minimisation G_c cannot easily be assumed to be positive definite. This is in contrast to unconstrained minimisation where the Hessian of the objective function is expected to be positive definite at the unconstrained minimum. Since $\hat{G}_k$ is taken to be positive definite in (4.2.76) as well as in $\hat{q}$ in (4.2.103), the condition $\|\hat{G}_{k_0} - G_c\| \le \bar{t}_2$ in Lemma (4.9), requiring $\hat{G}_{k_0}$ to be chosen arbitrarily close to G_c, is in general too restrictive.

The convergence of the sequence $\{\underline{x}_k\}$ generated by (4.2.76) with appropriate stepsize strategies has already been established provided $\hat{G}_k$, $k=0,1,\ldots$ is taken as any bounded positive definite matrix. An alternative algorithm, based on a positive definite matrix $\hat{G}_k^L$, part of which is explained by $\hat{G}_k$, may therefore be constructed. Consider adding a nonnegative term to the right hand of (4.2.107) . Hence

$$f(\underline{x}_{k+1}) - f(\underline{x}_k) \le \tau_k \hat{q}_k(\underline{x}_p) + \frac{\tau_k}{2} < \underline{x}_p - \underline{x}_k, \hat{G}_k^g(\underline{x}_p - \underline{x}_k) > + \int_0^1 (1-t) \quad <.,.> dt \qquad (4.2.142)$$

where $<.,.>$ is the inner product in (4.2.107) and $\hat{G}_k^g$ is a positive semidefinite matrix chosen such that

$$\hat{G}_k + \hat{G}_k^g$$

is positive definite. We thus define $\hat{G}_k^L$ by

$$\hat{G}_k^L = \hat{G}_k + \hat{G}_k^g . \qquad (4.2.143)$$

Using this definition (4.2.142) can be rewritten as

$$f(\underline{x}_{k+1}) - f(\underline{x}_k) \le \tau_k \hat{q}_k^L(\underline{x}_p) + \int_0^1 (1-t) \quad <.,.> dt \qquad (4.2.144)$$

where

$$\hat{q}_k^L(\underline{x}_p) = <\nabla f(\underline{x}_k), \underline{x}_p - \underline{x}_k> + \tfrac{1}{2} <\underline{x}_p - \underline{x}_k, \hat{G}_k^L(\underline{x}_p - \underline{x}_k)> .$$

Provided $\hat{G}_k^L$ is positive definite, reformulating the projection problem (4.2.76) as

$$\underline{x}_p = \arg\min \{ \| \underline{x} - \underline{x}_k + \alpha_k \hat{H}_k^L \nabla f(\underline{x}_k) \|_{\hat{G}_k^L} \mid \underline{x} \in R \} , \qquad (4.2.145)$$

where $\hat{H}_k^L = (\hat{G}_k^L)^{-1}$, does not alter the convergence of $\{\underline{x}_k\}$ to $\underline{x}_c$. Similarly, the stepsize strategy (4.2.103) can be replaced by

$$f(\underline{x}_{k+1}) - f(\underline{x}_k) \le c_1 \tau_k \hat{q}_k^L(\underline{x}_p) \ . \tag{4.2.146}$$

Provided $\hat{G}_k^L$ is positive definite, the positive definiteness restriction on $\hat{G}_k$ can be relaxed. Subject to the restriction (4.2.143), $\hat{G}_k$ can be chosen to be arbitrarily near G_c. Thus, following the same argument as in (4.2.107) - (4.2.110) of Theorem (4.9), a suitable choice of $\hat{G}_k$ ensures that the restriction

$$0 < c_1 \le 1 - \frac{\tau_k}{\ell\xi} \left[\frac{1}{2} \frac{||(G_c - \hat{G}_k)(\underline{x}_{k+1} - \underline{x}_k)||}{||\underline{x}_{k+1} - \underline{x}_k||} - \mu_k^c\right] < 1$$

or its simplified form (4.2.110) is satisfied with $\alpha_k = \tau_k = 1$ for some $k \ge K_o$, $K_o \ge o$, $t > o$, $\underline{x}_k \in \{\underline{x} \in E^n \mid ||\underline{x} - \underline{x}_c|| < t\}$. Hence whatever G_c may happen to be, if $\hat{G}_k^L$ is positive definite and large enough, $\hat{G}_k$ can be arbitrarily close to G_c and the choice $\alpha_k = \tau_k = 1$ is valid for the stepsize strategies at a close neighbourhood of $\underline{x}_c$. Given $\hat{G}_k^L$, the only restriction on $\hat{G}_k$ is that $\hat{G}_k^g = \hat{G}_k^L - \hat{G}_k$ should be positive semi-definite. A similar result can be derived for stepsize strategy (4.2.104), using (4.2.144) and the techniques in Theorem (4.9) with the above approach.

For positive definite $\hat{G}_k^L$, the convergence of $\underline{x}_k$ generated by subproblem (4.2.145),(4.2.1) and the steplength strategies (4.2.146) or (4.2.104) with $o < \alpha_k < 2$, follows from Theorem (4.8). Thus, as $\{\underline{x}_k\}$ converges to $\underline{x}_c$ and $\hat{G}_k$ can be arbitrarily close to G_c, (4.2.110) is satisfied for $\alpha_k = 1$ and τ_k converges to unity. Provided ℓ is taken to be $\ell||\underline{y}||^2 \le \langle\underline{y}, \hat{G}_k^L \underline{y}\rangle$, the results of Theorem (4.9) also remain valid for (4.2.145).

The above discussion assumes that $\underline{x}_k$, $\hat{G}_k$ are close enough to $\underline{x}_c$, G_c respectively so that unit steplengths, $\alpha_k = \tau_k = 1$ are achieved. It is hence desirable to show that if for $K_0 \geq o$, $\hat{G}_{K_0}$ is is chosen to be positive definite and in a close neighbourhood of G_c^L and if $\hat{G}_{K_0}$ satisfies (4.2.110), for some $\mu_{K_0}^c > o$ such that $\ell/2 - \mu_{K_0}^c > o$, hence $\tau_{K_0} = \alpha_{K_0} = 1$ with

$$\hat{G}_{K_0}^L = \hat{G}_{K_0} + \hat{G}_{K_0}^g$$

for a positive semi-definite $\hat{G}_{K_0}^g$, then $\hat{G}_k$, $k \geq K_0$ will remain in a close neighbourhood of $\hat{G}_c$. This ensures that $\tau_k = \alpha_k = 1$ for $k \geq K_0$. It should be noted that the requirement for $\underline{x}_{K_0}$ to be in a close neighbourhood of $\underline{x}_c$ is implicit in the assumption $\ell/2 - \mu_{K_0}^c > o$, because $\mu_{K_0}^c$ is given by (4.2.108) and because of continuity.

As required in Lemma (4.9), let $\|\underline{x}_{K_0} - \underline{x}_c\| \leq \bar{t}_1$ and let $\|\hat{G}_k - G_c\| \leq t_2$ for $k \geq K_0$. Using the same argument as in (4.2.107) - (4.2.110), $\alpha_k = \tau_k = 1$ can be concluded. For unit steplengths Theorem (4.9) shows that $\underline{x}_{k+1}$ computed using (4.2.145) satisfies $\|\underline{x}_{k+1} - \underline{x}_c\| \leq r^{k+1-K_0}\,\bar{t}_1$ for some $r \in (0,1), k \geq K_0$. This can be used in Lemma (4.9), with the R-linear convergence property established in Theorem (4.9), to show that $\|\hat{G}_k - G_c\| \leq t_2$ for updates satisfying (4.2.122),provided $\|\hat{G}_{K_0} - G_c\| \leq \bar{t}_2$ holds.The argument to establish $\|\hat{G}_{k+1} - G_c\| \leq t_2$ remains the same as in Lemma (4.9), using the results in Theorem (4.9) for subproblem (4.2.145). Thus, if (4.2.122) is satisfied, then $\|\underline{x}_{k+1} - \underline{x}_c\| \leq \bar{t}_1$, $\|\hat{G}_{k+1} - G_c\| \leq t_2$ and using the same arguments as above $\alpha_{k+1} = \tau_{k+1} = 1$ can be established. As this can be done for any $k \geq K_0$, $\|\underline{x}_k - \underline{x}_c\| \leq \bar{t}_1$, $\|\hat{G}_k - G_c\| \leq \bar{t}_2$, unit steplengths can be maintained if (4.2.122) is satisfied. To show that (4.2.122) is satisfied for subproblem (4.2.145) with unit steplengths and (4.2.128) with $\underline{c} = D\,\underline{\delta}$ or $\underline{c} = \underline{\delta}$ consider inequality (4.2.140) of Theorem (4.10). This condition has been derived without any assumptions on the projection subproblem used in the algorithm. Thus it equally applies to subproblem (4.2.145). The inequality $(1 - \lambda\,\theta^2)^{\frac{1}{2}} \le (1 - \frac{1}{2}\lambda\theta^2)$ may be used to reduce (4.2.140) to the required condition (4.2.122). It has already been shown above that the updating formula (4.2.128) with $\underline{c} = \underline{\delta}$ or $\underline{c} = D\,\underline{\delta}$ (D positive definite) satisfies condition (4.2.133), required for (4.2.140). Thus, if $\|\hat{G}_{K_0} - G_c\| \le \bar{t}_2$ and $\|\underline{x}_{K_0} - \underline{x}_c\| \le \bar{t}_1$, using the updating formula (4.2.128) for $\hat{G}_k$ and (4.2.143) for $\hat{G}^L_k$ and assuming that $\underline{x}_{K_0}$, $\underline{\lambda}_{K_0}$, $\hat{G}^L_{K_0}$ have been chosen to be in a close neighbourhood of $\underline{x}_c$, $\underline{\lambda}_c$, $\hat{G}^L_c$, then for $\underline{c} = D\,\underline{\delta}$ or $\underline{c} = \underline{\delta}$ $\hat{G}^L_{k+1} = \hat{G}_{k+1} + \hat{G}^g_{k+1}$ holds and $\hat{G}_k$ can be regarded conceptually as being computed by (4.2.128) with $\|\hat{G}_k - G_c\| \le t_2$, $k \ge K_0$. Hence $\hat{G}_k$ remains in a close neighbourhood of G_c and unit steplengths, $\alpha_k = \tau_k = 1$, are maintained in subproblem (4.2.145). Thus, superlinear convergence results for subproblem (4.2.76) apply to algorithms based on subproblem (4.2.145).

From Theorem (4.3) (see (4.2.44), (4.2.50) and Theorem (4.9) (see (4.2.109)), it follows that when α_k is near 2 in algorithms (4.1.10) and (4.1.19) with stepsize strategies (4.2.37) or (4.2.103) or (4.2.104) or (4.2.146) and $0 < \alpha_k \le \bar{\alpha} < 2$, τ_k may have to be set to a small value to satisfy conditions like (4.2.44),(4.2.50), (4.2.109). Thus, the computation of τ_k may take longer and the convergence of the sequence $\{\tau_k\}$ to unity may be slower than what it might be for lower values of α_k . It is therefore proposed that when τ_k is lower than a prescribed bound, the values of α_k are reduced in subsequent iterations. Lowering α_k to much less than unity may allow τ_k to converge rapidly to one and thereby induce savings in the computation of τ_k satisfying one of the above steplength criteria. However this tends to hinder the progress of the algorithm since α_k determines the length of the unconstrained step. Thus, setting α_k to much less than unity should also be avoided.

4.3 CONCLUDING REMARKS

In this chapter a projection algorithm for nonlinear programming is formulated. This algorithm can be used with a number of stepsize strategies without altering its convergence properties. When approximate Hessians are used, conditions derived in Section (4.2.3) show that the algorithm achieves unit steplengths in a neighbourhood of the optimum if the current approximation to the Hessian of $f(\underline{x})$ is near enough to the Hessian of $f(\underline{x})$ evaluated at the constrained optimum.

The algorithm discussed above requires all the successive points it generates to remain feasible. A useful extension of this algorithm would therefore be one which does not impose this restriction and computes projections on local linear approximations of the feasible region, converging simultaneously to a feasible and optimal solution. Such an extension is currently under consideration.

CHAPTER 5

THE ITERATIVE SPECIFICATION OF OBJECTIVE FUNCTIONS IN ECONOMIC POLICY OPTIMISATION: AN APPLICATION OF PROJECTION METHODS

5.1 INTRODUCTION

A fundamental problem in the optimisation of policy decisions is the specification of a suitable objective function. It has been argued by Rustem, Velupillai and Westcott (1978) and Westcott, Holly, Rustem and Zarrop (1976) that quadratic functions penalising the weighted deviation of the computed trajectories from their desired values form an acceptable class of objective functions. In this Chapter an iterative method is described for specifying the weighting matrices (i.e. the Hessian) of such quadratic functions. The approach is based on projection techniques and was first discussed in Rustem, Velupillai, Westcott (1978). This chapter also includes certain refinements of the original method. The method is not concerned with a "best" set of weights *independent* of a "best" desired path. A politically acceptable path, optimally generated, is the main aim. Since the method is a formalisation of possible (and necessary) interactions between the policy-maker and the model builder, starting with almost any set of initial weights, convergence is expected if the two parties are consistent in the successive iterations and a plausible (i.e. politically acceptable) set of desired paths is given. In fact, since the method is iterative, there is no loss of generality in beginning with identical weights. Once the initial "optimal path" has been determined the method provides a systematic way in which the policy-maker can state his dissatisfaction with the values of the various trajectories and tell the model builder the changes he would like to see, and leaves it to the model builder to update the weighting matrix to generate a more acceptable path. The ammendments to the optimal trajectory, to make it more acceptable, are translated into corresponding rank-one corrections to the weighting matrices. It is proposed that an "optimal set" of weights may be obtained by repeated updating so that, at the end of this iterative procedure, the "final" optimal trajectory will be totally acceptable. The iterative procedure is thus aimed at tackling the problem in a way which, for example, Kornai has described as follows: "It is in the course of planning that economic administration gains an increasingly concrete knowledge of what *can be done*, and comes, accordingly, to realise what it wants to be done. .... it is *in the course of planning* on the basis of information obtained, that the economico-political targets to be prescribed will become increasingly clear. This cognition process is not a steady one". (Kornai (1975, p. 420), cf. also, Sen (1970, p. 63)).

Basic concepts and the reformulation of the policy optimisation problem as a projection problem are given in Section (5.2). In Section (5.3) the method is described and its justification is established using projection techniques. In Section (5.4) the computational aspects and convergence of the method are discussed. Numerical results illustrating the method are given in the Appendix.

5.2 PRELIMINARIES

A quadratic objective function can be considered equivalent to the second order Taylor series expansion about the unconstrained optimum of the true nonlinear objective function. The constant term of this expansion can be ignored without affecting the value of the optimal trajectory. Consider, therefore, the quadratic function

$$J(\underline{Y},\underline{U}) = \tfrac{1}{2} < \left(\begin{bmatrix} \underline{Y} \\ \hline \underline{U} \end{bmatrix} - \begin{bmatrix} \underline{Y} \\ \hline \underline{U} \end{bmatrix}^d \right), G \left(\begin{bmatrix} \underline{Y} \\ \hline \underline{U} \end{bmatrix} - \begin{bmatrix} \underline{Y} \\ \hline \underline{U} \end{bmatrix}^d \right) > \qquad (5.2.1)$$

where G is a symmetric (n+m)K x (n+m)K matrix,

$$\underline{Y}^T \triangleq \left[\underline{y}^T(1),\ldots, \underline{y}^T(k),\ldots,\underline{y}^T(K)\right] \qquad (5.2.2)$$

is the mK-vector of the trajectory of the endogenous variables (outputs) in the time period $[1,K]$ with the m-vector $\underline{y}(k)$, $1 \leq k \leq K$, denoting the endogenous variables at time k,

$$\underline{U}^T \triangleq \left[\underline{u}^T(1),\ldots, \underline{u}^T(k),\ldots,\underline{u}^T(K)\right] \qquad (5.2.3)$$

is the nK-vector of the trajectory of policy instruments(controls) in the period $[1,K]$ with the n-vector $\underline{u}(k)$ denoting the policy instruments at time k. The superscript d on $\underline{Y}$ and $\underline{U}$ denotes the desired trajectories or targets. The desired trajectories are also the minima of the Taylor series expansion (5.2.1). Since this expansion is about the unconstrained minimum of the true nonlinear objective function, $\underline{Y}^d$ and $\underline{U}^d$ are exactly the values of $\underline{Y}$, $\underline{U}$ which minimise the true welfare function. The assumption that the true nonlinear objective function has a finite unconstrained minimum may be relaxed, if required, by choosing a subset of the equality constraints such that a finite minimum exists in their intersection. The desired trajectories may then be interpreted as this minimum. The vectors $\underline{Y}^d$ and $\underline{U}^d$ may be thought of as desired but possibly infeasible values considering the restriction imposed by the constraints of the system.

Definition (5.1)

The optimisation problem involved in the minimisation of (5.2.1) subject to the restrictions of the model equations as well as any other constraints is defined as

$$\min \{J(\underline{Y}, \underline{U}) \mid \underline{Y}, \underline{U} \in F, \underline{h}(\underline{Y},\underline{U}) = \underline{0}, \underline{g}(\underline{Y},\underline{U}) \geq \underline{0}\} . \qquad (5.2.4)$$

The vector valued function $\underline{h}$ denotes the model equations as well as any other equality constraints imposed on $\underline{Y}$ and $\underline{U}$. The vector function $\underline{g}$ denotes the inequality constraints on the values of $\underline{Y}$ and $\underline{U}$. The set constraint $F \subset E^{(n+m)K}$ is the set of admissible values of $\underline{Y}$, $\underline{U}$ from the policy maker's point of view.

Definition (5.2)

The set of $\underline{Y}$, $\underline{U}$ satisfying the functional constaints of (5.2.4) is defined as

$$R \triangleq \{\underline{Y}, \underline{U} \in E^{(n+m)K} \mid \underline{h}(\underline{Y}, \underline{U}) = \underline{0}, \underline{g}(\underline{Y}, \underline{U}) \geq \underline{0}\} . \qquad (5.2.5)$$

Remark (5.1)

Under the assumption of convexity of R (e.g. when $\underline{h}$ and $\underline{g}$ are linear) the solution to

$$\min \{J(\underline{Y}, \underline{U}) \mid \underline{Y}, \underline{U} \in R\} \qquad (5.2.6)$$

may be interpreted as the point on R nearest to $\underline{Y}^d$, $\underline{U}^d$ with respect to the norm

$$\left\| \begin{bmatrix} \underline{Y} \\ \hline \underline{U} \end{bmatrix} - \begin{bmatrix} \underline{Y} \\ \hline \underline{U} \end{bmatrix}^d \right\|_G^2 \qquad (5.2.7)$$

for positive definite G. Hence the solution to (5.2.6) is the projection of the desired trajectories $\underline{Y}^d$, $\underline{U}^d$ onto R. When G is singular $J(\underline{Y},\underline{U})$ has no finite unconstrained minimum. In this case a subset of the equality constraints may be chosen such that G is positive definite in their intersection. A finite minimum will hence exist on these constraints.

The relationship between the original desired trajectories and this minimum is defined through the chosen constraints. An equivalent problem to (5.2.6) may thus be formulated with a positive definite G in (5.2.7) and with the minimum in the intersection of the chosen constraints as the corresponding desired trajectories. In fact, the numerical example in the Appendix only assumes the part of G weighting the policy instruments (controls) to be positive definite. The sub-matrix weighting the endogenous variables (outputs) is allowed to be positive semi-definite. Thus, the matrix G is, in general, positive semi-definite.

Assumption (5.1)

The feasible set of the optimisation problem (5.2.4) is not empty, i.e. $R \cap F \neq \phi$. This simple assumption is central to the discussion in subsequent sections. Points in this intersection have to be sought by solving (5.2.6) for different $J(\underline{Y}, \underline{U})$. Thus (5.2.6) has to be solved a number of times by respecifying $J(\underline{Y}, \underline{U})$ until a solution $\left[\underline{Y}^T \,\vdots\, \underline{U}^T\right]^*$ is found such that

$$\begin{bmatrix} \underline{Y} \\ \hdashline \underline{U} \end{bmatrix}^* \in F \cap R \tag{5.2.8}$$

It should be noted that in contrast to the algebraic equations describing R, the subspace F exists only in the mind of the rational policy maker.

Remark (5.2)

Dynamic optimisation problems are included in formulation (5.2.4). The correspondence between dynamic and static optimisation problems are extensively discussed in Canon, Cullum, Polak (1970) and Polak (1971). Definitions (5.1) and (5.2) for $\underline{Y}$ and $\underline{U}$ have been made to suit such multi-time period decision problems.

Given the desired values $\left[\underline{Y}^T \,\vdots\, \underline{U}^T \right]^d$, and the region R which will be assumed to be convex from now on, the nearness of the solution of (5.2.6) to $\left[\underline{Y}^T \,\vdots\, \underline{U}^T \right]^d$ is affected only by the weighting matrix G in (5.2.1). Different values of this matrix will therefore define different points on R as the nearest point to $\left[\underline{Y}^T \,\vdots\, \underline{U}^T \right]^d$. Thus, given these desired trajectories the only way of respecifying $J(\underline{Y}, \underline{U})$ to produce an optimal solution in $F \cap R$ is to respecify G.

Let G_c denote the current weighting matrix of (5.2.1). The corresponding solution of (5.2.6) using G_c will be called "the current optimal solution",

$$\left[\underline{Y}^T \,\vdots\, \underline{U}^T \right]^*_c . \tag{5.2.9}$$

After a careful inspection of the values of (5.2.9), the policy maker may decide that some of these values are not quite what he wants them to be. This means (5.2.9) does not satisfy his set constraint, i.e.

$$\left[\underline{Y}^T \,\vdots\, \underline{U}^T \right]^*_c \notin F . \tag{5.2.10}$$

One way of obtaining an optimal solution which the policy maker may prefer to the current one (5.2.9) is to alter some of the elements of the weighting matrix G_c. The basic idea is to update G_c to a new weighting matrix denoted by G_n such that "the new optimal trajectory"

$$\left[\underline{Y}^T \,\vdots\, \underline{U}^T \right]^*_n \tag{5.2.11}$$

obtained by solving (5.2.6) with G_n is also in F (see Figure 1). The simple conjecture that the new optimal solution (5.2.11) may not be completely satisfactory (i.e. not in F) as a result of the first update of G leads to an iterative updating procedure. After each update, the optimisation problem is solved using the new G until a solution is obtained which is also in F.

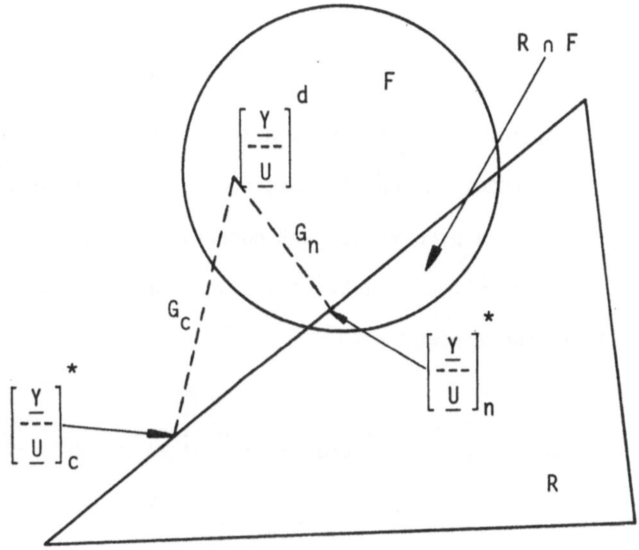

Figure 1

For the purpose of computing the updates of G, the policy maker is required to specify the trajectory he prefers to the current optimal solution (5.2.9). Hence the "preferred" trajectory

$$\left[\underline{Y}^T \mid \underline{U}^T\right]^P \tag{5.2.12}$$

incorporates all the corrections to (5.2.9) such that (5.2.12) is preferable to the "current" optimal trajectory. An obvious property of the preferred trajectory is that it is in F. The weighting matrix G has thus to be updated such that the next optimal solution (5.2.11) is in the neighbourhood of (5.2.12) (see Lemma (5.2)). Given the "current" optimal trajectory (5.2.9) and the "preferred" trajectory (5.2.12) the displacement (or correction) vector is defined as

$$\underline{\delta} \stackrel{\Delta}{=} \begin{bmatrix} \underline{Y} \\ --- \\ \underline{U} \end{bmatrix}^{P} - \begin{bmatrix} \underline{Y} \\ --- \\ \underline{U} \end{bmatrix}^{*}_{c} . \qquad (5.2.13)$$

Let $J_{true}(\underline{Y}, \underline{U})$ and $\nabla J_{true}(\underline{Y}, \underline{U})$ denote respectively the true non-linear objective function and its gradient. This true objective function need not be seen as a purely abstract concept. If it is taken to be a quadratic , its weighting is what is actually being sought (see Theorem (5.1)). The vector

$$\underline{\gamma} = \nabla J_{true}(\underline{Y}^{P}, \underline{U}^{P}) - \nabla J_{true}(\underline{Y}^{*}_{c}, \underline{U}^{*}_{c}) \qquad (5.2.14)$$

is thus the difference of the gradients of J_{true} evaluated at (5.2.12) and (5.2. 9). The weighting matrix which will give a new optimal solution (5.2.11) nearer to the preferred values may be computed by adding a rank-one correction term to the current weighting matrix using the following formula

$$G_n = G_c + \frac{(\underline{\gamma} - G_c\,\underline{\delta})\,(\underline{\gamma} - G_c\,\underline{\delta})^T}{\langle \underline{\gamma} - G_c\underline{\delta}\,,\,\underline{\delta} \rangle} \qquad (5.2.15)$$

This updating formula is a member of the variable metric family widely used for computing approximations to second derivative matrices in unconstrained (see e.g. Broyden (1967) and constrained optimisation (see Murtagh and Sargent (1969)). The vector $\underline{\gamma}$ represents the change in the gradient vector of the true objective function evaluated at the preferred and at the current optimal trajectories. Clearly, assigning values to $\underline{\gamma}$ is not a trivial problem and is discussed in Section (5.4).

The conceptual motivation underlying the choice of the rank-one formula will be given by Theorem (5.1). The actual computation of the updates when the value of $\underline{\gamma}$ is not exactly known involves a simplification of (5.2.15) to another formula (5.4.2). In Section (5.4), this simpler formula will be derived from the general formula of the variable metric family to which (5.2.15) belongs .

Remark

As the weighting matrix G is symmetric, the quadratic function (5.2.1) is said to penalise deviations of the optimal solution (5.2. 9) from the desired trajectories in a desirable direction as much as a deviation in an undesirable direction. For example, if the desired value for unemployment is 1.00% and the optimal solution is at .5%, the quadratic welfare function has penalised the (.5-1.0 = - .5%) departure as much as it would have if the optimal solution has been 1.5%. Clearly, such cases are the result of *misspecification* of the desired trajectories. The iterative nature of the method enables it to take account of this. If an optimal trajectory is equal to or higher (in the desirable direction) than its corresponding desired value in the desired trajectory, then a new optimisation criterion may be defined by resetting those desired values marginally higher than the current optimal value. Thus in the case of the unemployment example above, the desired value may be changed to some value less than .5%. This will eventually ensure that all departures from the desired trajectory are in undesirable directions. Hence the problems introduced by the symmetric nature of the quadratic criterion function should no longer be considered crucial. It should be noted, however, that such changes may sometimes be at the expense of the optimal value of some other variable. The problem of symmetry can also be treated using piecewise quadratic welfare functions and thereby introducing asymmetry (Friedman (1972)). The use of the updating formula (5.2.15) may also be extended to such functions. However, resetting the desired trajectories and thereby avoiding the problems due to symmetry seems to be a simpler approach than solving the problem after it has occurred in a rather complicated manner.

5.3 THE CONCEPTUAL ALGORITHM

Using (5.2.15) adjustments are made to G to generate optimal policies which the policy maker may prefer to the current optimal trajectory (5.2. 9). The process of adjusting the weights may be repeated until a satisfactory analysis is made and an acceptable optimal solution is attained. Figure 2 illustrates the iterative nature of the method. As in Figure 2, if the new optimal trajectory (5.2.11) is not quite in F, the procedure of adjusting G has to be repeated. Conceptually, the resulting iterative procedure needs to follow the following steps:

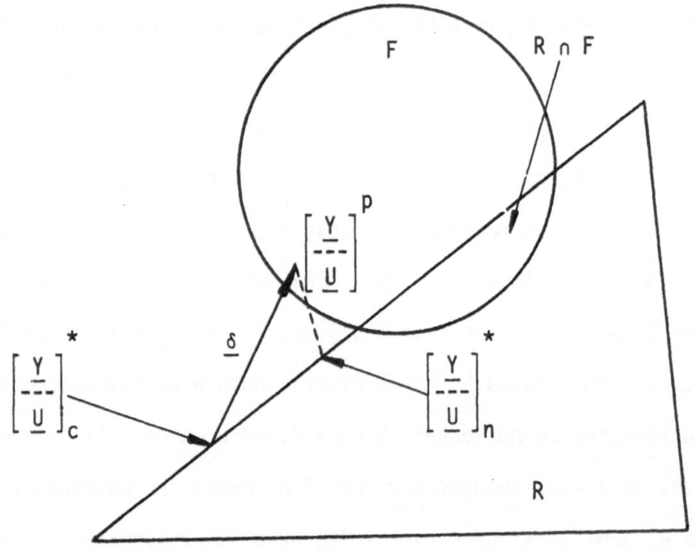

Figure 2

Algorithm (5.3.1)

Step 0:

Given the desired trajectory $\begin{bmatrix} \underline{Y} \\ \hline \underline{U} \end{bmatrix}^d$ assume some initial weighting matrix G_c.

Step 1:

Using the weighting matrix G_c and the desired trajectory in (5.2.1) solve the minimisation problem (5.2.6) to obtain the current optimal trajectory (5.2.9).

Step 2:

(This is the step summarising the role of the policy maker/economist in the respecification of G. He may have either formulated mathematical inequalities which the solution is expected to satisfy or just a feel for the values of the optimal trajectory). If, according to the policy maker

$$\begin{bmatrix} \underline{Y} \\ \hline \underline{U} \end{bmatrix}_c^* \in F \quad \text{(and hence} \quad \begin{bmatrix} \underline{Y} \\ \hline \underline{U} \end{bmatrix}_c^* \in R \cap F\text{) stop.}$$

Otherwise ask the policy maker to specify the changes required in the current optimal trajectory to make it acceptable. The preferred trajectory (5.2.12) is thus specified along with the corresponding $\underline{\Upsilon}$ vector (5.2.14). Methods for specifying $\underline{\Upsilon}$ are discussed in Section (5.4).

Step 3:

Given the vector $\underline{\Upsilon}$, the current optimal and the preferred trajectories compute the displacement vector $\underline{\delta}$ (5.2.13) and using the updating formula (5.2.15) compute G_n. Set $G_c = G_n$ and go to Step 1.

The correction computed using (5.2.15) in Step 3 has the interesting property that it does not restrict the choice of the direction or magnitude of $\underline{\delta}$ in any way. The denominator of (5.2.15) must be safeguarded from becoming zero. This can be accomplished with small changes to $\underline{\delta}$ and/or $\underline{\gamma}$ within limits acceptable to the policy maker. Provided this is done, G_n remains positive definite if G_c is positive definite.

Theorem (5.1) (Fiacco and McCormick (1968))

Motivation for Convergence

Assume the true objective function to be a positive definite quadractic form. Assume, also, that (n+m)K linearly independent $\underline{\delta}$ vectors and corresponding values for $\underline{\gamma}$ may be specified during the iterative procedure. If no zero denominators occur in the updating formula then starting with any positive definite initial weighting matrix, the generated updates converge to the weighting matrix of the true objective function after at most (n+m)K iterations. □

An important aspect of this result is that the specified $\underline{\delta}$ and $\underline{\gamma}$ vectors need to be consistent with the true welfare function. Theorem (5.1) remains as only a motivation for the convergence of the iterative procedure precisely because it may be very difficult for the policy maker to satisfy such requirements on $\underline{\delta}$ and $\underline{\gamma}$.

The effect of respecifying the weighting matrix on the optimal trajectories will be investigated under the assumption of only linear equality constraints describing the region R.

Definition (5.3)

The convex region R in (5.2.5) is assumed to be defined only by ℓ

linearly independent equations and is redefined as the linear manifold

$$R\,(\underline{b}) \triangleq \{\underline{Y}, \underline{U} \in E^{(n+m)K} \mid N^T \begin{bmatrix} \underline{Y} \\ \hline \underline{U} \end{bmatrix} = \underline{b}\} \qquad (5.3.2)$$

where N is an ℓ x (m+n)K matrix and $\underline{b}$ an ℓ-vector. The ℓ columns of N are thus linearly independent and $\ell \leq$ (m+n)K . The equations of the model are included in (5.3.2). The subspace of $R(\underline{b})$ is clearly $R(\underline{0})$.

As in Remark (5.2), the current optimal trajectory obtained with G_c in (5.2.6) is the projection of the desired trajectories onto $R(\underline{b})$ with respect to the norm (5.2.7) and is given by

$$\begin{bmatrix} \underline{Y} \\ \hline \underline{U} \end{bmatrix}_c^* = \begin{bmatrix} \underline{Y} \\ \hline \underline{U} \end{bmatrix}^d - H_c\, N\, (N^T H_c N)^{-1} \left(N^T \begin{bmatrix} \underline{Y} \\ \hline \underline{U} \end{bmatrix}^d - \underline{b} \right) \qquad (5.3.3)$$

where $H_c = G_c^{-1}$. Correspondingly, the projection operator

$$P = I - H_c N (N^T H_c N)^{-1} N^T \qquad (5.3.4)$$

projects all vectors in $E^{(n+m)K}$ onto the subspace $R(\underline{0})$, The derivations of (5.3.3) and (5.3.4) are given in Chapter 3.

Lemma (5.1)

If G_n is computed as the update of G_c using (5.2.15), the new optimal solution (5.2.11) may be expressed in terms of the current optimal solution and a correction term:

$$\begin{bmatrix} \underline{Y} \\ \hline \underline{U} \end{bmatrix}_n^* = \begin{bmatrix} \underline{Y} \\ \hline \underline{U} \end{bmatrix}_c^* + \alpha\, \underline{\beta} \qquad (5.3.5)$$

where the scalar

$$\alpha = \frac{(H_c \underline{\gamma} - \underline{\delta})^T G_c \left(\begin{bmatrix} \underline{Y} \\ \underline{U} \end{bmatrix}_c^* - \begin{bmatrix} \underline{Y} \\ \underline{U} \end{bmatrix}^d \right)}{(H_c \underline{\gamma} - \underline{\delta})^T N(N^T H_c N)^{-1} N^T (H_c \underline{\gamma} - \underline{\delta}) - (H_c \underline{\gamma} - \underline{\delta})^T \underline{\gamma}} \tag{5.3.6}$$

and the vector

$$\underline{\beta} = P(H_c \underline{\gamma} - \underline{\delta}) \tag{5.3.7}$$

is the projection of $(H_c \underline{\gamma} - \underline{\delta})$ onto $R(\underline{0})$. Thus the correction to the current optimal solution lies along this projection.

<u>Proof</u>

The new optimal trajectory is obtained by replacing G_c with G_n in (5.3.3). The Sherman-Morrison formula may be used with (5.2.15) to express H_n and $(N^T H_n N)^{-1}$ in (5.3.3) in terms of H_c and $(N^T H_c N)^{-1}$. After some rearranging

$$\begin{bmatrix} \underline{Y} \\ \underline{U} \end{bmatrix}_n^* = \begin{bmatrix} \underline{Y} \\ \underline{U} \end{bmatrix}_c^* - \frac{(H_c \underline{\gamma} - \underline{\delta})^T N(N^T H_c N)^{-1} \left(N^T \begin{bmatrix} \underline{Y} \\ \underline{U} \end{bmatrix}^d - \underline{b}\right)}{\Delta(\underline{\gamma}, \underline{\delta})} \underline{\beta} \tag{5.3.8}$$

is obtained where the scalar $\Delta(\underline{\gamma}, \underline{\delta})$ denotes the denominator of (5.3.6). The relationship

$$\nabla J(\underline{Y}_c^*, \underline{U}_c^*) = N \underline{\lambda} \tag{5.3.9}$$

is given by Lagrange multiplier theory for an ℓ-vector $\underline{\lambda}$. Since the current optimal trajectory is feasible and hence lies on R $(\underline{b})$, the equality

$$N^T \begin{bmatrix} \underline{Y} \\ \underline{U} \end{bmatrix}_c^* = \underline{b} \tag{5.3.10}$$

holds. Thus, using (5.3.10) and the fact that (5.3.9) denotes the gradient of the quadratic function (5.2.1)

$$N^T \begin{bmatrix} \underline{Y} \\ - \\ \underline{U} \end{bmatrix}^d - \underline{b} = - N^T \left(\begin{bmatrix} \underline{Y} \\ - \\ \underline{U} \end{bmatrix}_c^* - \begin{bmatrix} \underline{Y} \\ - \\ \underline{U} \end{bmatrix}^d \right) = - N^T H_c \nabla J(\underline{Y}_c^*, \underline{U}_c^*)$$

$$= - N^T H_c N \underline{\lambda} \qquad (5.3.11)$$

is obtained. Substituting $J(\underline{Y}_c^*, \underline{U}_c^*)$ in (5.3.11) and (5.3.9) in (5.3.8) gives the required result. □

5.4 COMPUTATIONAL CONSIDERATIONS FOR THE IMPLEMENTATION OF THE ALGORITHM

The main problem with the implementation of the algorithm discussed in Section (5.3) is the specification of $\underline{\gamma}$ (5.2.14). This vector measures the change in the gradient of the true objective function. The gradient with respect to the endogenous variable $y_i(k)$ or policy instrument $u_j(k)$ at time k evaluated at the fixed values $y_i^o(k)$, $u_j^o(k)$ denotes the "priority" attached to a unit increase from $y_i^o(k)$ or $u_j^o(k)$. Thus $\underline{\gamma}$ denotes the change in these "priorities" when going from the current optimal to the preferred trajectories. In Bray (1972, 1975) a way is described for quantifying these gradients by asking policy makers direct questions about their priorities on different variables. This is similar, in spirit, to the propositions made by Frisch (1972). If $\underline{\gamma}$ is obtained in this way, Theorem (5.1) ensures the convergence, in finite number of steps, of the algorithm in Section (5.3) to the true weighting matrix of the true quadractic objective function.

Apart from the difficulties of specifying a different $\underline{\gamma}$ at each iteration and of fulfilling the consistency requirement discussed in Section (5.3), a further complication arises when policy makers prefer not to be asked questions about their priorities on different variables. An alternative and more mechanical way of choosing $\underline{\gamma}$ has been developed to overcome such problems. The basis of this method lies in the desirable form the new optimal solution is expressed in. The vector $\underline{\gamma}$ is approximated with a scalar multiple of G_c, i.e.

$$\underline{\gamma} \approx \phi G_c \underline{\delta} \tag{5.4.1}$$

hence the updating formula (5.2.15) becomes

$$G_n = G_c + (\phi-1) \frac{G_c \underline{\delta} \underline{\delta}^T G_c}{< \underline{\delta}, G_c \underline{\delta} >} \tag{5.4.2}$$

Most well-known variable metric formulae belong to the family given by

$$G_n = G_c - \frac{G_c \underline{\delta} \underline{\delta}^T G_c}{< \underline{\delta}, G_c \underline{\delta} >} + \frac{\underline{\gamma}\underline{\gamma}^T}{< \underline{\delta}, \underline{\gamma} >} + \tau \underline{w}\underline{w}^T \tag{5.4.3}$$

where $\underline{w} = G_c \underline{\delta} - (<\underline{\delta}, G_c\underline{\delta}> / <\underline{\delta},\underline{\gamma}>)\underline{\gamma}$ and τ is a scalar. Different updating formulae are obtained by assigning different values to τ. The Davidon-Fletcher-Powell, Broyden-Fletcher-Goldfarb-Shanno and the rank-one (5.2.15) updating formulae may be obtained from (5.4.3) for corresponding values of τ (see, Dennis and Moré (1977), Powell (1974). It is interesting to note that when (5.4.1) is substituted into (5.4.3), (5.4.2) is obtained. Hence all variable metric formulae which may be derived from (5.4.3) reduce to (5.4.2) when (5.4.1) is used.

Lemma (5.2)

If $\underline{\gamma}$ is given by (5.4.1) and hence G_n by (5.4.2) the correction term in (5.3.5) reduces to a scalar multiple of the projection of $\underline{\gamma}$ onto $R(\underline{0})$. Thus the next optimal trajectory is given by

$$\begin{bmatrix} \underline{Y} \\ \hline \underline{U} \end{bmatrix}_n^* = \begin{bmatrix} \underline{Y} \\ \hline \underline{U} \end{bmatrix}_c^* + \alpha' P \underline{\delta} \qquad (5.4.4)$$

where P is the projection operator (5.3.4) with

$$\alpha' = - \frac{(\phi-1)\, \underline{\delta}^T H_c \left(\begin{bmatrix} \underline{Y} \\ \hline \underline{U} \end{bmatrix}_c^* - \begin{bmatrix} \underline{Y} \\ \hline \underline{U} \end{bmatrix}^d \right)}{\underline{\delta}^T N(N^T H_c N)^{-1} N^T \underline{\delta} + \phi (G_c \underline{\delta})^T H_c P(G_c \underline{\delta})}$$

If

$$\phi \geq 1 \qquad (5.4.5)$$

and if

$$\underline{\delta}^T G_c \left(\begin{bmatrix} \underline{Y} \\ \hline \underline{U} \end{bmatrix}_c^* - \begin{bmatrix} \underline{Y} \\ \hline \underline{U} \end{bmatrix}^d \right) \leq 0 \qquad (5.4.6)$$

then $\alpha' \geq 0$ and $\alpha' > 0$ when both (5.4.5) and (5.4.6) are strict inequalities.

<u>Proof</u>

Substituting (5.4.1) for $\underline{\gamma}$ into (5.3.8) yields (5.4.4) with

$$\alpha' = \frac{(\phi-1)\, \underline{\delta}^T H_c \left(\begin{bmatrix} \underline{Y} \\ \hline \underline{U} \end{bmatrix}_c^* - \begin{bmatrix} \underline{Y} \\ \hline \underline{U} \end{bmatrix}^d \right)}{(\phi-1)\, \underline{\delta}^T N(N^T H_c N)^{-1} N^T \underline{\delta} - \phi \underline{\delta}^T G_c \underline{\delta}} \qquad (5.4.7)$$

Using $H_c G_c \underline{\delta}$ instead of $\underline{\delta}$, the denominator of (5.4.7) may be expressed as

$$\Delta(\phi G_c \underline{\delta}, \underline{\delta}) = -\phi[(G_c \underline{\delta})^T H_c (G_c \underline{\delta}) - (G_c \underline{\delta})^T H_c N(N^T H_c N)^{-1} N^T H_c (G_c \underline{\delta})]$$

$$- \underline{\delta}^T N(N^T H_c N)^{-1} N^T \underline{\delta} = -\phi (G_c \underline{\delta})^T H_c P(G_c \underline{\delta}) - \underline{\delta}^T N(N^T H_c N)^{-1} N^T \underline{\delta}$$

where P is the projection operator (5.3.4). The result follows immediately since $H_c P$ is a symmetric operator, the columns of N are linearly independent and H_c is symmetric positive definite. □

The correction term in (5.4.4) , $\alpha' P\underline{\delta}$, is along the projection of $\underline{\delta}$ onto $R(\underline{0})$. This implies that the new optimal solution is as 'near' to the specified preferred solution as allowed by the model and the other constraints which define the manifold $R(\underline{b})$, defined by (5.3.2). The scalar α' has to be restricted and depends on the choice of ϕ and $\underline{\delta}$. The inequality (5.4.6) may be interpreted as a "rationality condition" on the choice of $\underline{\delta}$. The reason for this lies in the form of the new optimal solution (5.4.4). When (5.4.5) and (5.4.6) hold, $\alpha' \geq 0$ and thus the modification to the current optimal trajectory in (5.4.4) lies in the same direction as the projection of $\underline{\delta}$. This is the "best" alternative, in the sense of norm (5.2.7), allowed by the manifold $R(\underline{b})$ when $\underline{\delta}$ is infeasible and lies outside $R(\underline{b})$. Also, since $\underline{\delta}$ is defined by the preferred trajectory (5.2.12), it is a descent direction at the current optimal trajectory. As

$$\nabla J(\underline{Y}_c, U_c) = G_c\left(\begin{bmatrix} \underline{Y} \\ \hdashline \underline{U} \end{bmatrix}_c^* - \begin{bmatrix} \underline{Y} \\ \hdashline \underline{U} \end{bmatrix}^d\right)$$

condition (5.4.6) confirms that $\underline{\delta}$ is a descent direction. The following theorem establishes the convergence of the method when the new optimal solutions are expressed via (5.4.4). This is done by proving, for restricted α', that the new optimal solution (5.4.4) is nearer to the preferred trajectory than the current optimal solution.

<u>Theorem (5.2)</u> (Convergence)

If the new optimal trajectory is given by (5.4.4) with $0 < \alpha' \leq 2$, then the inequality

$$\left\|\begin{bmatrix}\underline{Y}\\ \hline \underline{U}\end{bmatrix}^{P} - \begin{bmatrix}\underline{Y}\\ \hline \underline{U}\end{bmatrix}^{*}_{c}\right\|_{G_c} \geq \left\|\begin{bmatrix}\underline{Y}\\ \hline \underline{U}\end{bmatrix}^{P} - \begin{bmatrix}\underline{Y}\\ \hline \underline{U}\end{bmatrix}^{*}_{n}\right\|_{G_c} \qquad (5.4.8)$$

holds. Furthermore, for $0 < \alpha' < 2$ strict inequality holds in (5.4.8).

Proof

Let the vector $\underline{p}$ denote, in short, the preferred trajectory (5.2.12), $\underline{c}$ denote the current optimal trajectory (5.2.9), $\underline{n}$ denote the new optimal trajectory (5.2.11) given by (5.4.4) and $\underline{q}$ denote the projection of $\underline{p}$ onto $R(\underline{b})$ which is also equivalent to the new optimal trajectory (5.4.4) when $\alpha' = 1$. Thus, $\underline{\delta} = \underline{p} - \underline{c}$, $\alpha' P\underline{\delta} = \underline{n} - \underline{c}$, $P\underline{\delta} = \underline{q} - \underline{c}$. In terms of these vectors, the required result (5.4.8) becomes

$$\|\underline{\delta}\|_{G_c} \geq \|\underline{p}-\underline{n}\|_{G_c}. \qquad (5.4.9)$$

Since $\underline{q}$ is the projection of $\underline{p}$, the inequality

$$< \underline{p}-\underline{q}, G_c(\underline{q}-\underline{c}) > \geq 0 \qquad (5.4.10)$$

follows from Lemma (4.2). Let the vector $\underline{w}(\alpha')$ denote points along $\underline{q}$-$\underline{c}$, starting from $\underline{c}$, thus

$$\begin{aligned}\underline{w}(\alpha') &= \alpha'\underline{q} + (1-\alpha')\underline{c}\\ &= \underline{c} + \alpha'(\underline{q} - \underline{c}), \quad 0 < \alpha' \leq 2 \qquad (5.4.11)\end{aligned}$$

Consider the function $f(\alpha') = < \underline{p} - \underline{w}(\alpha'), G_c(\underline{p} - \underline{w}(\alpha')) >$

Using (5.4.11) this can be written as

$$f(\alpha') = < \underline{p}-\underline{c}, G_c(\underline{p}-\underline{c}) - 2\alpha' < \underline{p}-\underline{c}, G_c(\underline{q}-\underline{c}) >$$

$$+ (\alpha')^2 < \underline{q} - \underline{c}, G_c(\underline{q} - \underline{c})> \quad (5.4.12)$$

From (5.4.10) it follows that

$$< \underline{p} - \underline{c}, G_c (\underline{q}-\underline{c}) > = <\underline{p} - \underline{q}, G_c(\underline{q}-\underline{c})> + < \underline{q}-\underline{c}, G_c(\underline{q} - \underline{c})>$$

$$\geqslant 0$$

and thus for $0 < \alpha' \leqslant 2$, the inequality

$$- 2 \alpha' < \underline{p} - \underline{c}, G_c (\underline{q}-\underline{c}) > + \alpha'^2 < \underline{q}-\underline{c}, G_c(\underline{q}-\underline{c})> \leqslant 0$$

holds. Thus $f(\alpha') \leqslant || \underline{\delta} ||^2_{G_c}$ and for $\underline{n} = \underline{w} (\alpha'), || \underline{p} - \underline{n} ||^2_{G_c} = f(\alpha')$ from which the result (5.4.9) follows. Furthermore, for $0 < \alpha' < 2$, $f(\alpha') < || \underline{\delta} ||^2_{G_c}$ and strict inequality holds in (5.4.8). □

In Rustem, Velupillai, Westcott (1978) the bound on α' is given as $0 < \alpha' \leqslant 1$. This was due to a "cautious" approach to policy changes. In theory, however, the bound in Theorem (5.2) is valid.

5.5 CONCLUSIONS

The difficulty about the specification of the correct weighting matrix of a quadratic objective function is the scarcity of information about it. The method described in this chapter treats this problem at two levels of information. The first is when gradient information about the true objective function is available. In this case Theorem (5.1) establishes convergence to the true weights of the quadratic function. At the second level this information is not available and only the preferred trajectories are specified. Updating the weighting matrix at this level is aimed only at producing a new optimal trajectory which is more acceptable to the policy maker than the current one (see Theorem (5.2). In both cases the effect of the method in the modification of the optimal trajectories has been investigated. The numerical results discussed in the Appendix illustrate the method.

APPENDIX: A NUMERICAL EXAMPLE

The method for respecifying the weighting matrix of the quadratic objective function described in this paper was tested using the econometric model of the United Kingdom and the optimal control program discussed in Holly et al (1979). A standard dynamic programming procedure was used to solve the optimisation problem (5.2.6). A particular feature of this dynamic programming algorithm is that the structure of the weighting matrix of the objective function (5.2.1) is restricted. This is discussed in detail in Section (A.1). To start the respecification procedure, the initial set of weights given in Table 1 were taken to be those discussed in Holly et al (1979 , Case II) . The optimal control (5.2.6) is performed from the first quarter of 1970 to the last quarter of 1974. Within this period, the exogenous variables not subject to control have been set to their historical values and conceptually form part of the constant vector $\underline{b}$ in (5.3.2). The weighting matrix is respecified three times using (5.4.2) and the rationale behind each of these is given in Sections A.2, A.3 and A.4. Only the optimal trajectories of those variables with non-zero weights are listed in the tables. The variables with zero weights are allowed to deviate from their desired values as much as is required, within reasonable limits, by the optimal solution. It follows that the subspace F of the policy maker is fairly tolerant for variables assigned zero weights. The essential problem is thus the determination of the set of non-zero weighted variables that are in $F \cap R$. In the three respecifications that were done for the numerical example, the variables with zero weighting always remained within F. The behaviour

Table 1: The desired trajectories and initial weights

ENDOGENOUS VARIABLES (OUTPUTS) : $1 \le k \le 20$

Desired trajectories

y_1 : %Δ unemployment ; $y_1^d(k) = \{0, -1, -1, -.5, 0, \ldots, 0\}$

y_{11} : %Δ exports ; $y_{11}^d(k) = \{1.75, 1.75, \ldots, 1.75\}$

y_{12} : %Δ imports ; $y_{12}^d(k) = \{1.5, 1.5, 1.5, \ldots, 1.5\}$

y_{18} : GDP deflator ; $y_{18}^d(k) = \{1.34, 1.34, \ldots, 1.34\}$

Weights

$a_{1,1}(k) = \{2.0, 2.0, \ldots, 2.0\}$,

$a_{11,11}(k) = \{2.0, 2.0, \ldots, 2.0\}$,

$a_{12,12}(k) = \{2.0, 2.0, \ldots, 2.0\}$,

$a_{18,18}(k) = \{2.0, 2.0, \ldots, 2.0\}$,

POLICY INSTRUMENTS (CONTROLS) : $1 \le k \le 20$

Desired trajectories

$u_1(k)$: %Δ government current expenditure ; $u_1^d(k) = \{.437, .437, \ldots, 437\}$

$u_2(k)$: change in the minimum deposit rate; $u_2^d(k) = \{0, 0, \ldots, 0\}$

$u_3(k)$: weighted exchange rate index ; $u_3^d(k) = \{0, 0, \ldots, 0\}$

$u_4(k)$: value added tax ; $u_4^d(k) = \{0, 0, \ldots, 0\}$

$u_5(k)$: cumulated discretionary indirect taxes ; $u_5^d(k) = \{0, 0, \ldots, 0\}$

$u_6(k)$: basic tax rate ; $u_6^d(k) = \{0, 0, \ldots, 0\}$

$u_7(k)$: aggregated personal allowances ; $u_7^d(k) = \{0, 0, \ldots, 0\}$

$u_8(k)$: national insurance contributions ; $u_8^d(k) = \{3.17, \ldots 3.17\}$

Table 1: (continued)

For each year of the Five Year Horizon

	1st Quarter	2nd Quarter	3rd Quarter	4th Quarter
$b_{1,1}$	1	1	1	1
$b_{2,2}$	3	3	3	3
$b_{3,3}$	.01	.01	.01	.01
$b_{4,4}$	3	3	3	3
$b_{5,5}$	3	3	3	3
$b_{6,6}$	15	.01	15	15
$b_{7,7}$	15	.01	15	15
$b_{8,8}$	1	1	1	1

(Note: % ≡ percentage, Δ ≡ change)

of the variables with non-zero weightings is described below.

The results reported in this section do not amount to an exhaustive attempt to produce a completely acceptable vector of trajectories since without the active participation of a policy maker who formulates policies in the real world a "completely acceptable" trajectory cannot be defined. Thus, the purpose of this section is the illustration of the method with a numerical example.

A.1 The Structure of the Objective Function

The objective function (5.2.1) used in the optimal control problem implicitly restricts the type of weighting allowed on the deviations of $\underline{y}(k)$ and $\underline{u}(k)$ from their desired values. This restriction prevents the weighting of the deviations at a given time with deviations at any other time. Thus deviations at time k of all the variables (endogenous variables and policy instruments) may only be weighted against all deviations at the same time. The structure for G that incorporates this constraint is the following:

$$G = \left[\begin{array}{ccc|ccc} A(1) & & & C^T(1) & & \\ & \ddots\ A(k) & & & \ddots\ C^T(k) & \\ & & \ddots\ A(K) & & & \ddots\ C^T(K) \\ \hline C(1) & & & B(1) & & \\ & \ddots\ C(k) & & & \ddots\ B(k) & \\ & & \ddots\ C(K) & & & \ddots\ B(K) \end{array}\right] .$$

The submatrices A(k), B(k) and C(k), $1 \leq k \leq K$, are the weights on the deviations of $\underline{y}(k)$, $\underline{u}(k)$ and $\underline{y}(k)$ against $\underline{u}(k)$, respectively. The matrices A(k) are positive semi-definite, whereas B(k) are strictly positive definite.

To preserve the above structure of G during an update (5.2.15), the preferred trajectory (5.2.12) is set equal to the current optimal trajectory (5.2.9) except at time k. At k, the modifications to the current optimal trajectory during this period are stored in the corresponding part of the preferred trajectory. Thus, the displacement vector $\underline{\delta}$ (5.2.13) is zero except at k. Using (5.4.1) to compute $\underline{Y}$ preserves the structure (5.4.1) during its respecification via (5.4.2) since this updating formula is used to update only the submatrices at k. For this purpose the matrix G in (5.4.2) is replaced by its submatrices at k, i.e.

$$\left[\begin{array}{c|c} A\ (k) & C^T(k) \\ \hline C\ (k) & B\ (k) \end{array}\right] .$$

The new weighting matrix G_n may thus be written with the new values of A(k), B(k) and C(k) replacing their respective current values. Hence, the updated G has the same structure as above and only those positions corresponding to time k are altered.

The updating formula (5.4.2) with $\underline{Y}$ given by (5.4.1) does not assign a nonzero weight to a variable which has not been weighted originally. Thus, variables with nonzero weights remain with such weights and those with zero weights remain zero weighted throughout the iterative respecification procedure.

A.2 First Respecification

The initial weights, the desired trajectories and the definitions of all the weighted variables are given in Table 1. The quadratic objective function was set up using these values. This function was then used as the objective of the optimisation problem (5.2.6). The current optimal trajectories obtained as the solution of (5.2.6) are given in Table 2.

Certain elements of the current optimal trajectory have undesirable values. Consider the second period of 1972, 1972(2), for example. Aggregate allowances u_7 are reduced by 4 per cent. It is difficult to imagine the actual reduction of the nominal value of personal allowances. This then is an ideal candidate for respecification. During 1972(2) imports (y_{12}) rise by 2.12 per cent, and prices (y_{18}) rise by 1.59 per cent (quarter on quarter). Both these rises are high and so they are included in the respecification.

For the reason discussed in Section (A.1), each respecification has to concentrate on one time period only. The second period of 1972 is chosen for this respecification. The specified preferred values are given in Table 3 along with the corresponding values for $\underline{\delta}$ and the weighting matrices respecified using (5.4.2) with $\phi = 3$.

Thus, y_{12} is required to move more towards its original desired value, y_{18} to actually go below its original desired value, while u_7 is required to move towards zero. The objective function was then updated and the optimal control program was rerun. The results showing the new optimal trajectories are listed in Table 4. Some improvement has taken place. While u_7 seems to move very readily

	ENDOGENOUS VARIABLES				POLICY INSTRUMENTS							
	y_1	y_{11}	y_{12}	y_{18}	u_1	u_2	u_3	u_4	u_5	u_6	u_7	u_8
70(1)	2.7068	-.0661	2.4947	1.2606	4.6358	-.1472	-1.9200	-.0093	-.0257	-.0252	.0191	-.2355
70(2)	-.0582	1.9629	2.3969	.0248	1.6295	-.0243	.4852	.0006	.0017	-6.2487	4.7199	-.0389
70(3)	.1760	-.7611	.6272	1.1677	.8440	-.0155	-1.9800	-.0023	-.0063	-.0027	.0020	-.0248
70(4)	-.6972	2.9965	1.5588	.5793	-.5592	.0676	.6229	.0029	.0080	.0116	-.0087	.1081
71(1)	-.5907	.8643	1.2230	.8889	.1766	.0381	-.7763	-.0019	-.0052	.0065	-.0049	.0610
71(2)	.0564	1.5026	1.2381	.9433	.9954	.0001	.6162	-.0018	-.0050	.0241	-.0182	.0002
71(3)	.4824	.9104	1.3397	.9728	.7052	.0174	-3.2890	-.0036	-.0100	.0030	-.0023	.0279
71(4)	-.1045	.9240	1.7308	.7377	.6584	.0237	5.6179	.0040	.0112	.0041	.0031	.0379
72(1)	.2694	1.1065	.7829	1.1354	.9035	.0117	-.5773	-.0075	-.0208	.0020	-.0015	.0188
72(2)	-.3146	1.3569	2.1216	1.5895	1.0093	.0206	-.2146	.0029	.0080	5.3003	-4.0036	.0330
72(3)	.4684	1.0025	.5718	1.0020	1.2801	-.0061	-2.8207	-.0094	-.0261	-.0010	.0008	-.0098
72(4)	-.0881	1.1362	2.2144	.8757	.5428	.0444	1.1611	.0049	.0136	.0076	-.0058	.0710
73(1)	.0629	.3390	1.0859	.5586	-.0340	.0723	-1.0318	-.0136	-.0378	.0124	-.0094	.1157
73(2)	.3559	1.7527	1.4210	1.1029	1.1504	1.9088	.3652	.0312	.0865	-1.2805	.9672	1.4399
73(3)	.1975	.9236	1.4084	.9794	.6394	.0182	-2.6745	-.0036	-.0100	.0031	-.0024	.0291
73(4)	.1674	2.1642	1.2204	1.3497	.6341	.0105	.3265	-.0037	-.0104	.0018	-.0014	.0169
74(1)	-.4679	.8369	1.2509	1.4129	.7597	-.0005	9.8481	-.0047	-.0129	0	0	-.0008
74(2)	.2616	1.7305	1.4837	2.6847	1.2943	-.0311	-12.5547	-.0033	-.0091	-7.9951	6.0391	-.0497
74(3)	.5457	1.0779	1.3223	1.8600	.5402	-.0053	0	-.0025	-.0070	-.0009	.0007	-.0085
74(4)	-.4549	1.1274	1.7764	1.9329	.0288	.0212	0	.0015	.0043	.0036	-.0027	.0339

Table 2: The initial optimal trajectories.

Table 3: Details of the First Respecification

VARIABLES AT (1972(2))	PREFERRED VALUES: Only those different from the current optimal avlues are listed, the rest are set equal to their current optimal value	CURRENT OPTIMAL VALUES: declared undesirable by the policy maker; at (1972(2))	δ (Only the non-zero elements are listed)
y_{12}	1.5	2.1216	- .6216
y_{18}	1.2	1.5895	- .3895
u_7	0.0	- 4.0036	4.0036

Weights

Current Weights at 1972(2), for other quarters see Table 1.

A: endogenous variables:

$a_{1,1} = 2.0 \quad a_{11,11} = 2.0 \quad a_{12,12} = 2.0 \quad a_{18,18} = 2.0$

B: policy instruments:

$b_{1,1} = 1.0 \quad b_{2,2} = 3.0 \quad b_{3,3} = .01 \quad b_{4,4} = 3.0 \quad b_{5,5} = 3.0$

$b_{6,6} = .01 \quad b_{7,7} = .01 \quad b_{8,8} = 1.0$

All other elements of matrices A and B and whole of matrix C are set to zero.

New Weights (with $\underline{\gamma} = \phi\, Q_C \underline{\delta}$, $\phi = 3$)

A: endogenous variables:

$a_{1,1} = 2.0 \quad a_{11,11} = 2.0 \quad a_{12,12} = 4.4998 \quad a_{18,18} = 2.9817$

$a_{18,12} = a_{12,18} = 1.5665$

B: policy instruments:

$b_{1,1} = 1.0 \quad b_{2,2} = 3.0 \quad b_{3,3} = .01 \quad b_{4,4} = 3.0 \quad b_{5,5} = 3.0$

$b_{6,6} = .01 \quad b_{7,7} = .0126 \quad b_{8,8} = 1.0$

C: y - u weights:

$c_{7,12} = -.0805 \quad c_{7,18} = -.0505$

	ENDOGENOUS VARIABLES				POLICY INSTRUMENTS							
	y_1	y_{11}	y_{12}	y_{18}	u_1	u_2	u_3	u_4	u_5	u_6	u_7	u_8
70(1)	2.7087	-.0661	2.4941	1.2610	4.6321	-.1472	-1.8902	-.0093	-.0258	-.0253	.0191	-.2357
70(2)	-.0573	1.9629	2.3987	.0245	1.6253	-.0245	.5035	.0007	.0018	-6.2935	4.7538	-.0392
70(3)	.1770	-.7611	.6284	1.1682	.8356	-.0154	-1.8890	-.0022	-.0062	-.0026	.0020	-.0246
70(4)	-.6981	2.9965	1.5604	,5769	-.5613	.0674	.6016	.0031	.0086	.0115	-.0087	.1078
71(1)	-.5900	.8502	1.2277	.8853	.1605	.0392	-.7290	-.0021	-.0057	.0067	-.0051	.0627
71(2)	.0589	1.4939	1.2395	.9305	.9933	.0000	.6860	-.0017	-.0047	.0092	-.0070	.0001
71(3)	.5067	.8669	1.3375	.9628	.6828	.0186	-3.3179	-.0042	-.0116	.0032	-.0024	.0297
71(4)	-.1196	.9342	1.7523	.7045	.7077	.0202	5.3653	.0043	.0119	.0035	-.0026	.0323
72(1)	.2215	1.0839	.7973	1.0989	.9227	.0100	-.5859	-.0091	-.0252	.0017	-.0013	.0159
72(2)	-.2682	1.3235	2.0637	1.5770	.9612	.0334	-.2516	.0044	.0123	8.5786	-.5925	.0534
72(3)	.5318	1.0163	.5751	.9839	1.3119	-.0104	-2.7998	-.0093	-.0259	-.0018	.0013	-.0166
72(4)	-.0843	1.2569	2.2366	.8521	.4969	.0439	1.2186	.0052	.0145	.0075	-.0057	.0702
73(1)	.0506	.3427	1.0840	.5551	-.0700	.0711	-1.0421	-.0133	-.0371	.0122	-.0092	.1137
73(2)	.3384	1.7704	1.4212	1.1018	1.1325	1.9064	.3651	.0317	.0880	-1.2810	.9676	1.4399
73(3)	.1483	.9136	1.4082	.9804	.6348	.0183	-2.6770	-.0036	-.0099	.0031	-.0024	.0293
73(4)	.1692	2.1367	1.2162	1.3511	.6430	.0100	.3246	-.0038	-.0104	.0017	-.0013	.0160
74(1)	-.4648	.8418	1.2528	1.4095	.7623	-.0006	9.8298	-.0047	-.0129	-.0001	0	-.0010
74(2)	.2624	1.7305	1.4851	2.6805	1.2939	-.0311	-12.5605	-.0033	-.0091	-7.9912	6.0361	-.0497
74(3)	.5460	1.0791	1.3215	1.8566	.5399	-.0053	0	-.0025	-.0070	-.0009	.0007	-.0085
74(4)	-.4550	1.1283	1.7763	1.9321	.0285	.0212	0	.0015	-.0043	.0036	-.0027	.0339

Table 4: New optimal trajectories obtained via the first respecification at 1972(2), $\phi = 3$. All variables in %Δ.

towards its preferred value, the values for both y_{12} and y_{18} hardly change at all. This is clearly due to the relative weightings of policy instruments being higher than those for the endogenous variables. Since u_7 has changed it might be wondered where this instrument's impact makes itself felt. The closest analogue of u_7, given the form of the cost function, is u_6, the basic tax rate. The basic tax rate, to compensate for the constraint upon u_7 going negative, rises by 8.58 per cent rather than 5.30 per cent.

A.3 Second Respecification

The result of the second optimisation, listed in Table 4, has an undesirable value for u_1 in 1970(1). This is an increase in government current expenditure by 4.63 per cent in the first quarter of the control period (after the first respecification). As compared with the historical behaviour of government expenditure this represents a very rapid and probably infeasible increase during one single quarter. It was therefore decided that the preferred value for u_1 should be - 1.0 per cent (see Table 5). The result of doing this is shown in Table 6 for all the weighted instruments. The trajectories here should be compared with the corresponding values of Table 4. Since government current expenditure is such an important control, it would have a major impact upon the configuration of targets and instruments. The rate of increase of u_1 in period 1970(1) is reduced to 2.58 per cent, and the reduction is compensated for by part of the increased government expenditure being shifted to the second quarter. Some of the work formerly done by u_1 is now taken on by u_6 and u_7. Basic tax rate is cut by 11.42 per cent rather than 6.29 per cent; while allowances are increased by 8.63 per cent as with 4.75 per cent previously.

Table 5: Details of the Second Respecification

VARIABLE AT (1970(1))	PREFERRED VALUES Only those different from the current optimal values are listed, the rest are set equal to their current optimal value	CURRENT OPTIMAL VALUES declared undesirable by the policy maker; at (1970(1))	$\underline{\delta}$ (Only the non-zero elements are listed)
u_1	- 1.0	4.6321	- 5.6321

Weights

Current Weights at (1970(1), for other quarters see the new weights in Table 3 and Table 1.

A: Endogenous variables:

$a_{1,1} = 2.0 \quad a_{11,11} = 2.0 \quad a_{12,12} = 20 \quad a_{18,18} = 20$

B: Policy instruments:

$b_{1,1} = 1.0 \quad b_{2,2} = 3.0 \quad b_{3,3} = .01 \quad b_{4,4} = 3.0$

$b_{5,5} = 3.0 \quad b_{6,6} = 15.0 \quad b_{7,7} = 15.0 \quad b_{8,8} = 1.0$

All other elements of A,B and the whole of matrix C are zero.

New Weights (with $\underline{\gamma} = \phi\, Q_C \underline{\delta}$, $\phi = 3$)

A: Endogenous variables:
Same as the current weights.

B: Policy instruments:
$b_{1,1} = 3.0$; the rest are the same as the current weights.

C: y - u weights:
Null matrix.

	ENDOGENOUS VARIABLES				POLICY INSTRUMENTS							
	y_1	y_{11}	y_{12}	y_{18}	u_1	u_2	u_3	u_4	u_5	u_6	u_7	u_8
70(1)	3.7145	-.0661	2.1421	1.5018	2.5827	-.2408	-1.6984	-.0156	-.0432	-.0413	.0312	-.3853
70(2)	.5995	1.9629	2.7211	-.0794	2.3145	-.0444	.6417	-.0013	-.0036	-11.4206	8.6265	-.0711
70(3)	.3254	-.7611	.7579	1.1690	.9257	-.0145	-1.7995	-.0026	-.0071	-.0025	.0019	-.0232
70(4)	-.6913	2.9965	1.4761	.5990	-.5565	.0679	.6733	.0026	.0073	.0116	-.0088	.1086
71(1)	-.6158	.7585	1.2124	.9531	.1536	.0407	-.6961	-.0022	-.0061	.0070	-.0053	.0652
71(2)	.0603	1.4278	1.2060	.9762	1.0003	.0004	.7008	-.0020	-.0054	.0976	-.0737	.0006
71(3)	.5024	.8241	1.3457	-.0076	.6714	.0201	-3.3088	-.0042	-.0117	.0035	-.0026	.0322
71(4)	-.1177	.8999	1.7409	.7182	.6933	.0216	5.3792	.0042	.0116	.0037	-.0028	.0345
72(1)	.2136	1.0682	.8012	1.1019	.9053	.0118	-.6017	-.0091	-.0252	.0020	-.0015	.0188
72(2)	-.2690	1.3164	2.0488	1.5758	.9530	.0344	-.2846	.0043	.0120	8.8568	-.8590	.0551
72(3)	.5267	1.0119	.5777	.9767	1.2958	-.0084	-2.8351	-.0093	-.0259	-.0014	.0011	-.0135
72(4)	-.0882	1.2502	2.2346	.8434	.4802	.0460	1.1700	.0052	.0143	.0079	-.0060	.0737
73(1)	.0475	.3507	1.0812	.5443	-.0768	.0729	-1.0407	-.0134	-.0371	.0125	-.0094	.1166
73(2)	.3382	1.7862	1.4173	1.0935	1.1217	1.9021	.3653	.0313	.0869	-1.2773	.9648	1.4400
73(3)	.1846	.9305	1.4067	.9777	.6258	.0187	-2.6730	-.0036	-.0099	.0032	-.0024	.0300
73(4)	.1646	2.1600	1.2164	1.3512	.6360	.0104	.3277	-.0038	-.0104	.0018	-.0013	.0166
74(1)	-.4672	.8412	1.2499	1.4135	.7628	-.0006	9.8436	-.0047	-.0130	-.0001	0	-.0010
74(2)	.2629	1.7305	1.4831	2.6837	1.2955	-.0312	-12.5656	-.0033	-.0091	-8.0121	6.0519	-.0498
74(3)	.5465	1.0772	1.3213	1.8591	.5399	-.0053	0	-.0025	-.0070	-.0009	.0007	-.0085
74(4)	-.4558	1.1269	1.7756	1.9328	.0280	.0212	0	.0015	.0043	.0036	-.0027	.0339

Table 6: New optimal trajectories obtained via the second respecification at 1970(1), $\phi = 3$. All variables in %Δ.

An interesting consequence of this simple respecification is for unemployment. The increase in unemployment is not checked until the 4th quarter though by the second year the shape of the trajectory is very similar to that previously obtained. The trajectories of the remaining targets are hardly altered at all.

A.4 Third Respecification

Following the respecification in Table 5 and the optimal trajectory thereby obtained (see Table 6), once more the second quarter of 1972 is considered. This period was also the subject of the first respecification. Though u_7 was reduced markedly it still has a negative value. Equally, y_{12} and y_{18} did not move very far towards their preferred values. To insist that u_7 be positive its new preferred value is set at 3.5 per cent. At the same time the preferred value of y_{18} is reduced from 1.2 to 1.0 per cent. The preferred value for y_{12} remains at 1.5 per cent. Also, the government current expenditure, u_1, in 1970(2) is preferred to increase by 0.3 per cent rather than .95 (see Table 7). The setting of the value of ϕ can be used in this context to put additional pressure on y_{12} and y_{18} to move towards their preferred values. Two examples are given: the first with $\phi = 3$ and the second with $\phi = 10$ with corresponding respecifications given in Table 7. The new optimal trajectories obtained using the respecific-ation with $\phi = 3$ are given in Table 8. Compared with the result of the second respecification, the values for y_{12} and y_{18} are again little changed. The effect of this respecification on u_7 is once more marked. From a value of -.86 it increases to 4.92. To compensate for this u_6 goes from 8.86 to 12.00 per cent. Clearly the policy maker has to decide the trade-off between increases in allowances and increases in

Table 7: Details of the Third Respecification

VARIABLES AT (1972(2))	PREFERRED VALUES Only those different from the current optimal values are listed, the rest are set equal to their current optimal value	CURRENT OPTIMAL VALUES declared undesirable by the policy maker; at (1972(2))	$\underline{\delta}$ (only the non-zero elements are listed)
y_{12}	1.5	2.0488	- .5488
y_{18}	1.0	1.5758	- .5758
u_1	.3	.9530	- .6530
u_7	3.5	- .8590	4.359

WEIGHTS

Current Weights at 1972(2) are the new weights computed in Table 3. For other quarters see Tables 1 and 5.

New Weights (with $\underline{\gamma} = \phi Q_c \underline{\delta}$)

$\phi = 3$

A: Endogenous variables:

$a_{1,1} = 2.0$ $a_{11,11} = 2.0$ $a_{12,12} = 10.4750$ $a_{18,12} = a_{12,18} = 6.0555$

$a_{18,18} = 6.3541$

B: Policy instruments:

$b_{1,1} = 1.1839$ $b_{2,2} = 3.0$ $b_{3,3} = .01$ $b_{4,4} = 3.0$ $b_{5,5} = 3.0$ $b_{6,6} = .01$

$b_{7,7} = .0197$ $b_{1,7} = -.0361$ $b_{8,8} = 1.0$

C: y-u weights:

$c_{1,12} = 1.0481$ $c_{1,18} = .7874$ $c_{7,12} = -.2862$ $c_{7,18} = -.2050$

$\phi = 10$

A: Endogenous variables:

$a_{1,1} = 2.0$ $a_{11,11} = 2.0$ $a_{12,12} = 31.3885$ $a_{18,12} = a_{12,18} = 21.7669$

$a_{18,18} = 18.1573$

Table 7: (continued)

B: Policy instruments:

$b_{1,1} = 1.8274$ $b_{2,2} = 3.0$ $b_{3,3} = .01$ $b_{4,4} = 3.0$ $b_{5,5} = 3.0$

$b_{6,6} = .01$ $b_{7,7} = .0445$ $b_{7,1} = b_{1,7} = -.1623$ $b_{8,8} = 1.0$

C: y-u weights:

$c_{1,12} = 4.7167$ $c_{1,18} = 3.5434$ $c_{7,12} = -1.0059$ $c_{7,18} = -.7457$

All non-specified elements of A, B and C are set to zero.

	ENDOGENOUS VARIABLES				POLICY INSTRUMENTS							
	y_1	y_{11}	y_{12}	y_{18}	u_1	u_2	u_3	u_4	u_5	u_6	u_7	u_8
70(1)	3.7159	-.0661	2.1416	1.5022	2.5798	-.2405	-1.6253	-.0156	-.0434	.0412	-.0311	-.3848
70(2)	.6052	1.9629	2.7194	-.0784	2.3085	-.0444	.7011	-.0012	-.0034	-11.4123	8.6203	-.0710
70(3)	.3318	-.7611	.7612	1.1691	.9091	-.0137	-1.6286	-.0025	-.0070	-.0024	.0018	-.0219
70(4)	-.6912	2.9965	1.4821	.5914	-.5668	.0686	.6972	.0029	.0082	.0118	-.0089	.1097
71(1)	-.6158	.7236	1.2227	.9398	.1305	.0431	-.6128	-.0024	-.0067	.0074	-.0056	.0689
71(2)	.0682	1.3994	1.2057	.9449	1.0028	.0009	.7329	-.0018	-.0051	.2329	-.1759	.0015
71(3)	.5391	.7425	1.3486	.9646	.6455	.0226	-3.3332	-.0050	-.0140	.0039	-.0029	.0362
71(4)	-.1345	.8885	1.7727	.6429	.7705	.0175	5.0625	.0044	.0123	.0030	-.0023	.0281
72(1)	.1228	1.0283	.8361	1.0155	.9721	.0087	-.6179	-.0114	-.0316	.0015	-.0011	.0139
72(2)	-.1896	1.3011	2.0073	1.5383	.4777	.0465	-.3496	.0061	.0170	11.9667	4.9242	.0745
72(3)	.6331	1.0236	.5856	.9319	1.3772	-.0159	-2.8192	-.0092	-.0256	-.0027	.0021	-.0254
72(4)	-.0703	1.4016	2.2680	.7974	.4312	.0448	1.2338	.0056	.0155	.0077	-.0058	.0717
73(1)	.0365	.3584	1.0790	.5307	-.1249	.0712	-1.0562	-.0130	-.0361	.0122	-.0092	.1139
73(2)	.3156	1.8172	1.4174	1.0874	1.0910	1.8983	.3646	.0320	.0889	-1.2762	.9640	1.4418
73(3)	.1659	.9229	1.4061	.9773	.6115	.0192	-2.6769	-.0035	-.0097	.0033	-.0025	.0307
73(4)	.1646	2.1295	1.2109	1.3524	.6413	.0100	.3252	-.0038	-.0104	.0017	-.0013	.0160
74(1)	-.4650	.8486	1.2520	1.4091	.7630	-.0007	9.8178	-.0047	-.0129	-.0001	0	-.0011
74(2)	.2633	1.7308	1.4842	2.6786	1.2939	-.0311	-12.5742	-.0033	-.0090	-7.9975	6.0409	-.0498
74(3)	.5467	1.0790	1.3204	1.8552	.5300	-.0053	0	-.0025	-.0070	-.0009	.0007	-.0085
74(4)	-.4562	1.1280	1.7755	1.9323	.0276	.0212	0	.0015	.0043	.0036	-.0027	.0340

Table 8: New optimal trajectories obtained via the third respecification at 1972(2), $\phi = 3$. All variables in $\%\Delta$.

basic tax rate and this he can do by using the respecification method. The increase in government current expenditure (u_1) in 1972(2) is reduced from .95 to .48 but as with the second respecification this is compensated for by changes in government current expenditure in neighbouring quarters.

The optimal trajectories obtained using the respecification with $\phi = 10$ are given in Table 9. The variables y_{12} and y_{18} have been more successfully forced towards their preferred values. However, both u_7 and u_1 overshoot their preferred trajectories with consequent offsetting changes in u_6 and in u_1 in the neighbourhood of 1972(2). Since this overshooting was not objectionable the respecification was terminated. If it were desired to rectify this overshooting, the same run could be repeated with $\phi = 10$ and lower preferred values for u_1, u_7 and the same preferred values for y_{12} and y_{18}.

The iterative respecification need not stop here. The policy maker may decide to explore the consequences of modifying parts of the new optimal trajectory. This will also tell him much about the nature of the economic model and the kinds of hard economic constraints that face him when decisions have to be made.

	ENDOGENOUS VARIABLES				POLICY INSTRUMENTS							
	y_1	y_{11}	y_{12}	y_{18}	u_1	u_2	u_3	u_4	u_5	u_6	u_7	u_8
70(1)	3.7171	-.0661	2.1411	1.5025	2.5775	-.2402	-1.5645	-.0157	-.0435	-.0412	-.0311	-.3843
70(2)	.6099	1.9629	2.7180	-.0776	2.3036	-.0443	.7506	-.0012	-.0033	-11.4055	8.6151	-.0710
70(3)	.3372	-.7611	.7639	1.1692	.8953	-.0131	-1.4865	-.0025	-.0068	-.0022	.0017	-.0209
70(4)	-.6910	2.9965	1.4872	.5851	.5754	.0691	.7171	.0032	.0089	.0119	-.0090	.1106
71(1)	-.6159	.6946	1.2313	.9287	.1113	.0450	-.5435	-.0026	-.0072	.0077	-.0058	.0720
71(2)	.0743	1.3758	1.2055	.9189	1.0048	.0013	.7596	-.0017	-.0048	.3455	-.2609	.0022
71(3)	.5695	.6746	1.3510	.9331	.6239	.0247	-3.3534	-.0057	-.0159	.0042	-.0032	.0395
71(4)	-.1485	.8790	1.7992	.5803	.8346	.0142	4.7991	.0046	.0128	.0024	-.0018	.0227
72(1)	.0473	.9952	.8651	.9436	1.0277	.0062	-.6313	-.0133	-.0370	.0011	-.0008	.0099
72(2)	-.1236	1.2823	1.9727	1.5070	.0825	.0566	-.4036	.0076	.0212	14.5522	9.7322	.0905
72(3)	.7216	1.0332	.5921	.8947	1.4449	-.0221	-2.8060	-.0092	-.0254	-.0038	.0029	-.0353
72(4)	-.0553	1.5275	2.2957	.7592	.3905	.0438	1.2868	.0059	.0164	.0075	-.0057	.0701
73(1)	.0274	.3648	1.0772	.5194	-.1649	.0698	-1.0691	-.0127	-.0353	.0120	-.0090	.1117
73(2)	.2969	1.8430	1.4176	1.0824	1.0655	1.8952	.3640	.0326	.0906	-1.2753	.9633	1.4433
73(3)	.1503	.9165	1.4056	.9770	.5996	.1959	-2.6801	-.0035	-.0096	.0034	-.0025	.0313
73(4)	.1647	2.1041	1.2062	1.3533	.6458	.0096	.3231	-.0038	-.0104	.0017	-.0012	.0154
74(1)	-.4632	.8547	1.2538	1.4055	.7632	-.0007	9.7963	-.0046	-.0129	-.0001	0	-.0011
74(2)	.2637	1.7311	1.4851	2.6745	1.2926	-.0311	-12.5813	-.0033	-.0090	-7.9854	6.0317	-.0497
74(3)	.5468	1.0806	1.3196	1.8519	.5381	-.0053	0	-.0025	-.0070	-.0009	.0007	-.0085
74(4)	-.4565	1.1290	1.7753	1.9319	.0273	.0212	0	.0015	.0043	.0036	-.0028	.0340

Table 9: New optimal trajectories obtained via the third respecification at 1972(2), $\phi = 10$. All variables in %Δ.

CHAPTER 6

POLICY OPTIMISATION ALGORITHMS FOR NONLINEAR ECONOMETRIC MODELS

6.1 INTRODUCTION

In this chapter policy optimisation algorithms for nonlinear models are discussed. The approach taken in these algorithms is an extension of the projection methods in Section (1.1.4) to nonlinear equality constraints. In policy optimisation with econometric models the variables are naturally partitioned into independent (controls or policy instruments) and dependent (endogenous) variables. Thus, as in (1.1.77) - (1.1.78), exploiting this partition leads to the elimination of the dependent variables. This approach is based on the algorithms discussed in Rustem (1980).

As econometric models are systems of dynamic relationships, policy optimisation algorithms have sometimes been cast in dynamic optimisation terms and their solutions formulated using dynamic programming (see e.g. Bray (1975), Chow (1975), Holly et. al. (1979), Kendrick and Majors (1975), Wall and Westcott (1975)). In this chapter a static formulation of the policy optimisation problem has been adopted because of the nature of the discussion in subsequent sections. It is well known (see e.g. Davis (1977)) that in a deterministic framework (i.e. deterministic econometric model and objective function) such a static optimisation approach is equivalent to the dynamic programming formulation.

In policy optimisation the objective function is chosen to reflect the aims and preferences of the policy maker. A quadratic function has often been considered sufficient for this purpose (see Chapter 5 and Bray (1975), Chow (1975), Holbrook (1974), Kendrick and Majors (1975), Preston et al (1976), Wall and Westcott (1975). Hence adopting the terminology initially formulated in Section (5.2), consider the quadratic objective function

$$J(\underline{Y},\underline{U}) = \tfrac{1}{2}\langle \underline{Y} - \underline{Y}^d,\ G_y(\underline{Y} - \underline{Y}^d)\rangle + \tfrac{1}{2}\langle \underline{U} - \underline{U}^d,\ G_u(\underline{U} - \underline{U}^d)\rangle \tag{6.1.1}$$

where $\underline{Y}$ is the mK dimensional vector of endogenous variables (outputs) given by (5.2.2), $\underline{U}$ is the nK dimensional vector of policy instruments (controls) given by (5.2.3). The superscript d denotes desired values. The symmetric matrices G_y, G_u are the diagonal submatrices of G in (5.2.1), weighting $\underline{Y} - \underline{Y}^d$ and $\underline{U} - \underline{U}^d$ respectively. The off-diagonal submatrices of G in (5.2.1), weighting Y - U cross terms have been ignored in (6.1.1). The matrix G_y is assumed to be positive semidefinite and G_u is assumed to be strictly positive definite. The diagonal elements of these matrices penalise the departure of a variable from its desired value. The off-diagonal elements measure the importance attached to the deviation in one variable versus the deviation in another. Clearly, when both these matrices are positive definite, the desired trajectories $\underline{Y}^d$, $\underline{U}^d$ are also the minimum of the quadratic objective function. Although it is essential that the desired values are known exactly, in the absence of further information about the true nonlinear objective function of the policy maker, the matrices G_y and G_u may be determined using the iterative algorithm in Chapter 5 that involves interactions with the policy maker.

The policy optimisation problem may be formulated as the choice of $\underline{Y}$ and $\underline{U}$ such that the objective function (6.1.1) is minimised subject to the equations of the nonlinear econometric model throughout the period [1, K]. If these equations are written in compact form as $\underline{F}(\underline{Y}, \underline{U}) = \underline{0}$, the corresponding problem becomes

$$\min \{J(\underline{Y}, \underline{U}) \mid \underline{F}(\underline{Y}, \underline{U}) = \underline{0}\} . \tag{6.1.2}$$

The values of all the exogenous variables not subject to optimisation are assumed to have been substituted in $\underline{F}(\underline{Y}, \underline{U})$ before solving (6.1.2). Assuming positive definite G_y, G_u, (6.1.2) may be interpreted as computing a $\underline{Y}$, $\underline{U}$ pair that is as near as possible to $\underline{Y}^d, \underline{U}^d$. Thus, (6.1.2) is the nonlinear equality constrained extension of the projection problem discussed in (5.2.6). Finally, inequality constraints will not be considered since using the method discussed in Chapter 5, G_y and G_u may be adjusted to alter the solution of (6.1.2) such that the optimal values of $\underline{Y}, \underline{U}$ are acceptable to the policy maker.

In Section (6.2) the conventional simulations approach to policy making is shown to be a special case of (6.1.2). In Section (6.3) a Newton-type algorithm and its Gauss-Newton type alternative for solving (6.1.2) are discussed. In Section (6.4) quasi-Newton extensions to the Gauss-Newton algorithm are described. Numerical results are discussed in an appendix.

6.2 POLICY OPTIMISATION AND SIMULATIONS

The use of optimisation algorithms for policy decisions has mostly been evaluated against a background of the policy simulations approach (see, Committee on Policy Optimisation (1978)). Policy simulations basically consist of assuming values for the policy instruments, $\underline{U}^*$ say, and computing the corresponding set of endogenous variables, say $\underline{Y}^*$, using the econometric model. Thus, $\underline{F}(\underline{Y}^*, \underline{U}^*) = \underline{0}$ and if any part of $\underline{Y}^*, \underline{U}^*$ is unsatisfactory, another set of $\underline{U}$ may be specified for use in the subsequent simulation.

The simulation of the econometric model given $\underline{U}^*$ is equivalent to solving the simple optimisation problem

$$\min \{ J_u(\underline{U}) \mid \underline{F}(\underline{Y}, \underline{U}) = \underline{0} \} \qquad (6.2.1)$$

where $J_u(\underline{U})$ is a nonlinear function with a unique minimum at $\underline{U}^*$. This can be shown by writing the first order necessary conditions of optimality for (6.2.1). Thus, forming the Lagrangian

$$L(\underline{Y}, \underline{U}, \underline{\lambda}) = J_u(\underline{U}) + \underline{F}^T(\underline{Y}, \underline{U})\underline{\lambda}$$

where $\underline{\lambda}$ is the vector of Lagrange multipliers, the conditions

$$\frac{\partial L}{\partial \underline{U}} = \frac{\partial J_u}{\partial \underline{U}} + \left[\frac{\partial \underline{F}}{\partial \underline{U}}\right]^T \underline{\lambda} = \underline{0} \qquad (6.2.2,a)$$

$$\frac{\partial L}{\partial \underline{Y}} = \left[\frac{\partial \underline{F}}{\partial \underline{Y}}\right]^T \underline{\lambda} = \underline{0} \qquad (6.2.2,b)$$

$$\underline{F}(\underline{Y}, \underline{U}) = \underline{0} \qquad (6.2.2,c)$$

are obtained. Furthermore, (6.6.2,c) implies

$$\frac{\partial \underline{F}}{\partial \underline{U}} + \frac{\partial \underline{F}}{\partial \underline{Y}} \frac{\partial \underline{Y}}{\partial \underline{U}} = 0 . \qquad (6.2.3)$$

Using (6.2.3) with (6.2.2,a) yields

$$\frac{\partial J_u}{\partial \underline{U}} - \left[\frac{\partial \underline{Y}}{\partial \underline{U}}\right]^T \left[\frac{\partial \underline{F}}{\partial \underline{Y}}\right]^T = \underline{0}$$

which, in view of (6.2.2,b), reduces to

$$\frac{\partial J_u}{\partial \underline{U}} = \underline{0} \,. \tag{6.2.4}$$

Hence, the optimality conditions (6.2.2) reduce to solving (6.2.4) simultaneously with the model equations, (6.2.2,c). Since (6.2.4) is the necessary condition for the unconstrained minimum of $J_u(\underline{U})$, solving (6.2.4) and (6.2.2,c) simultaneously is equivalent to computing the unconstrained minimum, $\underline{U}^*$, satisfying (6.2.4) and then computing the corresponding $\underline{Y}^*$ from (6.2.2,c), given $\underline{U}^*$. Thus, the simulation of the model (6.2.2,c) with $\underline{U}^*$ given is equivalent to solving (6.2.1). □

In contrast to $J(\underline{Y}, \underline{U})$ in (6.1.1), the objective function $J_u(\underline{U})$ reflects the preferences of the policy maker in the policy variables only. In (6.2.1) the endogenous variables are allowed to take the values dictated by $\underline{U}^*$ and the model equations (6.2.2,c). In (6.1.2), the values $\underline{Y}$ may take are clearly influenced by the preferences reflected in $J(\underline{Y}, \underline{U})$. Thus, the solution of (6.1.2) has the advantage that the optimal value of $\underline{Y}$ may be controlled by $J(\underline{Y}, \underline{U})$. However, this cannot be accomplished by $J_u(\underline{U})$ in (6.2.1). In this sense the optimisation problem (6.2.1) may be seen as a special case of (6.1.2). Therefore, adopting (6.1.2) as a useful approach to policy decisions is justified. This leads us to consider efficient ways of solving (6.1.2). In subsequent sections a class of Newton and Quasi-Newton algorithms for solving (6.1.2) will be discussed.

6.3 A NEWTON-TYPE ALGORITHM

Large nonlinear econometric models are complex systems with which optimisation algorithms are allowed limited interaction. This is understandable given the size and complexity of the present macro-economic models and the fact that their structure is constantly altered by model builders (see, e.g. The London Business School Quarterly Econometric Model (August 1975/January 1979)). However, the following assumption is always valid.

Assumption (6.1)

Given $\underline{U} \in E^{nK}$, the set of equations $\underline{F}(\underline{Y}, \underline{U}) = \underline{0}$ may be solved for $\underline{Y} \in E^{mK}$ where E^{nK} and E^{mK} are respectively nK and mK dimensional Euclidean spaces. Hence, there exists a mapping g, provided by the solution program of an econometric model which computes $\underline{Y}$, given $\underline{U}$ such that

$$\underline{Y} = \underline{g}(\underline{U}) . \qquad (6.3.1)$$

The mapping (6.3.1) may be used to eliminate $\underline{Y}$ from (6.1.1). Thus

$$G(\underline{U}) \triangleq J(\underline{g}(\underline{U}), \underline{U}) = \tfrac{1}{2}\langle \underline{g}(\underline{U}) - \underline{Y}^d, \ G_y(\underline{g}(\underline{U}) - \underline{Y}^d)\rangle + \tfrac{1}{2}\langle \underline{U} - \underline{U}^d, G_u(\underline{U} - \underline{U}^d)\rangle, \qquad (6.3.2)$$

where the first term on the right hand side may be seen as a weighted least squares term. The original constrained problem (6.1.2) may be formulated as an equivalent unconstrained optimisation problem in $\underline{U}$ only, i.e.

$$\min\{ G(\underline{U}) \mid \underline{U} \in E^{nK} \} . \qquad (6.3.3)$$

Starting from an initial value of $\underline{U}$, say $\underline{U}^o$, the damped Newton method is a well known (see Ortega and Rheinboldt (1970)) iterative method for solving (6.3.3). The damped Newton method is based on the iteration

$$\underline{U}_{k+1} = \underline{U}_k + \alpha_k \underline{d}_k , \qquad k = 0,1,\ldots, \tag{6.3.4}$$

with

$$\underline{d}_k = - H_k \nabla G(\underline{U}_k). \tag{6.3.5}$$

In (6.3.4) the scalar $\alpha_k \geq 0$ is chosen to ensure that $G(\underline{U}_{k+1}) \leq G(\underline{U}_k)$ thereby enlarging the region of convergence of the Newton method (i.e. (6.3.4) with $\alpha_k \equiv 1$). In (6.3.5) $\nabla G(\underline{U}_k)$ denotes the gradient of $G(\underline{U})$ evaluated at $\underline{U}_k$, given by

$$\nabla G(\underline{U}_k) = N_k^T G_y(\underline{g}(\underline{U}_k) - \underline{Y}^d) + G_u(\underline{U}_k - \underline{U}^d) \tag{6.3.6}$$

where

$$N_k \triangleq \left.\frac{\partial \underline{g}}{\partial \underline{U}}\right|_{\underline{U} = \underline{U}_k} = \left.\frac{\partial \underline{Y}}{\partial \underline{U}}\right|_{\underline{U} = \underline{U}_k} \quad \text{with} \quad \frac{\partial Y_i}{\partial U_j} = \left[\frac{\partial \underline{Y}}{\partial \underline{U}}\right]_{i,j} , \tag{6.3.7}$$

and H_k denotes the inverse of the Hessian of $G(\underline{U})$ at $\underline{U}_k$, given by

$$G_k = N_k^T G_y N_k + B_k + G_u = H_k^{-1} \tag{6.3.8}$$

with

$$B_k = \sum_{i=1}^{mK} \nabla^2 g_i(\underline{U}_k)[G_y(\underline{g}(\underline{U}_k) - \underline{Y}^d)]_i$$

where i denotes the i th element of the corresponding vector and $\nabla^2 g_i(\underline{U}_k)$ is the Hessian of g_i with respect to $\underline{U}$, evaluated at $\underline{U}_k$. The second derivative term B_k in (6.3.8) is generally not available for econometric models and ignoring it leads to a generalised damped Gauss-Newton type method (see Ortega and Rheinboldt (1970)). The iteration for this method can be written as

$$\underline{U}_{k+1} = \underline{U}_k + \alpha_k \underline{d}_k^o , \qquad k = 0,1, \ldots, \qquad (6.3.9)$$

with

$$\underline{d}_k^o = - H_k^o \; \nabla G(\underline{U}_k) \qquad \text{and} \quad H_k^o = (G_k^o)^{-1} = N_k^T G_y N_k + G_u . \qquad (6.3.10)$$

As $G_y \geq 0$ and $G_u > 0$, G_k^o is positive definite and $\underline{d}_k^o$ is a descent direction. Thus the singularity problems that in general occur when B_k is ignored (see Gill and Murray (1976)), do not arise when $G_k^o > 0$ is guaranteed as in (6.3.10). Furthermore, it may be argued that econometric models are not very nonlinear so that ignoring B_k is justified since $||B_k||$ will be small compared to $||G_k^o||$. Given a feasible initial trajectory, $\underline{Y}_o$, $\underline{U}_o$, an algorithm based on (6.3.9) - (6.3.10) involves successive computations of the descent direction $\underline{d}_k^o$, the step size α_k and a convergence test to check whether the last value of $\underline{U}_k$ solves the optimisation problem (6.3.3).

<u>Algorithm</u> (6.3.11)

<u>Step 0:</u> Given $\underline{Y}^d$, $\underline{U}^d$, G_y, G_u and an initial trajectory $\underline{Y}_o$, $\underline{U}_o$ such that $\underline{F}(\underline{Y}_o, \underline{U}_o) = \underline{0}$ is satisfied, set $k = 0$.

<u>Step 1:</u> Compute $G(\underline{U}_k)$, (6.3.2); $\nabla G(\underline{U}_k)$, (6.3.6); N_k, (6.3.7) and H_k^o, (6.3.10).

<u>Step 2:</u> If the pair $\underline{Y}_k$, $\underline{U}_k$ is an adequate approximation to a minimum the algorithm is terminated. Gill and Murray (1976) discuss suitable convergence criteria for this purpose.

<u>Step 3:</u> Compute the descent direction $\underline{d}_k^o$ given by (6.3.10).

<u>Step 4:</u> Compute the steplength factor α_k such that $G(\underline{U}_{k+1})$ is

sufficiently lower than $G(\underline{U}_k)$ (see Ortega and Rheinboldt (1970)). Thus, compute $\underline{U}_{k+1}$ using (6.3.9) such that the inequality

$$G(\underline{U}_k) - G(\underline{U}_{k+1}) \geq -10^{-4}\alpha_k < \nabla G(\underline{U}_k), \underline{d}_k^o > / \|\nabla G(\underline{U}_k)\| \, \|\underline{d}_k^o\| \tag{6.3.12}$$

is satisfied.

Step 5: Compute $\underline{Y}_{k+1} = \underline{g}(\underline{U}_{k+1})$ using the solution program of the econometric model. Set $k = k+1$, go to Step 1.

For determining α_k in univariate searches as in Step 4, Gill and Murray (1974) suggest using safeguarded cubic or quadratic minimisation to compute $\bar{\alpha}$ as the approximate minimum of $G(\underline{U}_k + \alpha \underline{d}_k^o)$ along $\underline{d}_k^o$. Then α_k is set equal to the first member of the sequence $\{\bar{\alpha}(\frac{1}{2})^j;\ j = 0,1,2,\ldots\}$ that satisfies (6.3.12). As the computation of $G(\underline{U}_k + \alpha \underline{d}_k^o)$, for various values of α, involves the evaluation of $\underline{Y}$ using the model and $\partial\underline{Y}/\partial\underline{U}$ at $\underline{U}_k + \alpha\, \underline{d}_k$, cubic interpolation is very expensive for Algorithm (6.3.11). Even quadratic interpolation which only involves the evaluation of $\underline{Y}$ using the model solution program may be computationally very expensive. The Goldstein-Armijo (Ortega and Rheinboldt (1970)) procedure of choosing $\bar{\alpha} \geq -10^{-4} < \nabla G(\underline{U}_k), \underline{d}_k^o > / \|\underline{d}_k^o\|^2$ and setting α_k to the first member of the sequence $\{\bar{\alpha}(.1)^j;\ j = 0,1,2\ \ldots\}$ satisfying (6.3.12) was found to be satisfactory in numerical experiments. The simpler approach of setting $\bar{\alpha} = 1$ was also found satisfactory in practice.

The relation of (6.3.11) with the generalised Gauss-Newton Algorithm can be easily established if the optimisation problem (6.3.3) is interpreted as the solution of $(nK + mK)$ equations

$$\underline{g}(\underline{U}) - \underline{Y}^d = \underline{0}\,, \qquad \underline{U} - \underline{U}^d = \underline{0} \tag{6.3.13}$$

in nK variables. Clearly, (6.3.13) is an overdetermined system. Rewriting $G(\underline{U})$ as

$$G(\underline{U}) = \tfrac{1}{2}\|\underline{g}(\underline{U}) - \underline{Y}^d\|^2_{G_y} + \tfrac{1}{2}\|\underline{U} - \underline{U}^d\|^2_{G_u} \qquad (6.3.14,a)$$

where $\|\underline{z}\|^2_G \triangleq \langle \underline{z}, G\,\underline{z} \rangle$, the Gauss-Newton algorithm for solving (6.3.13) arises naturally if given $\underline{U}_k$, $\underline{g}(\underline{U}_k)$ and N_k,

$$G(\underline{U}) = \tfrac{1}{2}\|\underline{g}(\underline{U}) - \underline{g}(\underline{U}_k) + \underline{g}(\underline{U}_k) - \underline{Y}^d\|^2_{G_y} + \tfrac{1}{2}\|\underline{U} - \underline{U}_k + \underline{U}_k - \underline{U}^d\|^2_{G_u} \qquad (6.3.14,b)$$

$\underline{U}_{k+1}$ is chosen as the value of $\underline{U}$ minimising the function

$$G_k(\underline{U}) = \tfrac{1}{2}\|N_k(\underline{U} - \underline{U}_k) + \underline{g}(\underline{U}_k) - \underline{Y}^d\|^2_{G_y} + \tfrac{1}{2}\|\underline{U} - \underline{U}_k + \underline{U}_k - \underline{U}^d\|^2_{G_u} \qquad (6.3.14,c)$$

obtained by linearising (6.3.13) about $\underline{U}_k$.[1] The result is given by (6.3.9) - (6.3.10) with $\alpha_k \equiv 1$. When $G_y = G_u = I$ the generalised Gauss-Newton method above reduces to the Gauss-Newton method. Thus, the basic convergence results related to the Gauss-Newton method apply to Algorithm (6.3.11). The convergence of the sequence $\{\underline{U}_k ;\ k = 0,1,2,\ldots\}$ computed using (6.3.9) - (6.3.10) follows from the damped Gauss-Newton theorem (Ortega and Rheinboldt (1970, Theorem 14.4.4)). The rate of convergence of the Gauss-Newton algorithm is at least linear (Ortega and Rheinboldt (1970, Theorem 14.4.6)) and this provides the convergence rate bound for (6.3.9) - (6.3.10) with $\alpha_k \equiv 1$. When $\underline{F}(\underline{Y}^d, \underline{U}^d) = \underline{0}$, the desired values happen to be feasible so that the equations (6.3.13) are consistent, the convergence rate for (6.3.9) - (6.3.10) with $\alpha_k \equiv 1$ is quadratic (see, e.g. Dennis (1977)). The convergence and superlinear rate of convergence of the damped Newton algorithm (6.3.4) follows from the damped Newton theorem (Ortega and Rheinboldt (1970, Theorem 14.4.3))

---

1 See footnote on page 276.

for twice continously differentiable $G(\underline{U})$. For $G(\underline{U})$ three times continously differentiable, quadratic rate of convergence is established by Goldstein (1967). However, due to the prohibitive computational requirements involved in evaluating the second derivative term B_k in (6.3.8), the damped Newton algorithm does not seem practicable for policy optimisation studies with large and nonlinear models.

Algorithm (6.3.11) is designed to be used with econometric models and their solution programs. As stated above, the solution program provides the mapping $\underline{g}$ in (6.3.1). This approach minimises the interaction between the optimisation program and the complex structure of the econometric model. The implementation of the algorithm is thereby simplified. An alternative is a Newton type approach to the original problem (6.1.2) that converges simultaneously to a feasible and optimal solution. However, such an approach would require the evaluation of $\partial\underline{F}/\partial\underline{Y}$ and $\partial\underline{F}/\partial\underline{U}$ at each $\underline{Y}_k$, $\underline{U}_k$. The difficulty in evaluating these gradients arises from the continuous modifications done by the model builders to the econometric model and from the complexity of the models. The modifications are done to update the model with incoming new statistical data. Updating the model would also require updating expressions for $\partial\underline{F}/\partial\underline{Y}$ and $\partial\underline{F}/\partial\underline{U}$. This has been a major source of error for model solution programs based on Newton's method. Evidently, the numerical evaluation of these derivatives is possible (see, Chow (1975), Mantell and Lasdon (1978)) but would impose heavy computational requirements for large models. Thus, Algorithm (6.3.11) that uses the mapping $\underline{Y} = \underline{g}(\underline{U})$ seems to be a suitable approach that avoids these difficulties.

Algorithm (6.3.11) requires the evaluation of N_k, $k = 0,1,\ldots,$ for each k. As the size of the nonlinear econometric model increases, so does the cost of computing N_k. One solution to this problem is to use and update an approximation to the matrix N_k instead of actually computing it. This approach will be described in the next section. A second-best approach is to devise a simplified Gauss-Newton method. In the neighbourhood of the optimum, N_k changes only slightly so that eventually, not changing N_k is an obvious tactic (see, e.g. Kowalik and Osborne (1968)). Assuming that the initial trajectory $\underset{\sim}{Y}_o, \underset{\sim}{U}_o$ in Algorithm (6.3.11) is reasonably close to the optimum, N_o may be used for all values of N_k. Thus, N is computed only once at $\underset{\sim}{Y}_o, \underset{\sim}{U}_o$. Such assumptions are not uncommon in optimisation studies with econometric models. Preston et al (1976) claim that only one computation of N for the Warton model has been sufficient. Chow (1979) claims that only three gradient evaluations have been sufficient for optimal control of the Michigan model. Hence, Algorithm (6.3.11) may be simplified by computing in Step 0, N_o and

$$G_o^o = N_o^T G_y N_o + G_u \tag{6.3.15}$$

instead of H_k^o. In Step 1 the gradient is computed using

$$\nabla G_o(\underset{\sim}{U}_k) = N_o^T G_y(\underline{g}(\underset{\sim}{U}_k) - \underline{Y}^d) + G_u(\underset{\sim}{U}_k - \underline{U}^d) . \tag{6.3.16}$$

The descent direction in Step 3 is computed using (6.3.15) - (6.3.16),

$$\hat{\underset{\sim}{d}}_k^o = - H_o^o \nabla G_o(\underset{\sim}{U}_k) \tag{6.3.17}$$

where $H_o^o = (G_o^o)^{-1}$, $\underset{\sim}{U}_{k+1}$ in Step 4 is given by

$$\underset{\sim}{U}_{k+1} = \underset{\sim}{U}_k + \alpha_k \hat{\underset{\sim}{d}}_k^o . \tag{6.3.18}$$

The convergence of the simplified Gauss-Newton algorithm (6.3.15) - (6.3.18) with $\alpha_k \equiv 1$ is proved in the following theorem.

Theorem (6.1)

Assume that $\nabla G_0 : D \subset E^{nK} \to E^{nK}$ is Frechet-differentiable on a convex set $D_0 \subset D$ and that

$$\| \nabla^2 G_0(\underline{U}_j) - \nabla^2 G_0(\underline{U}_t) \| \leq \gamma \| \underline{U}_j - \underline{U}_t \| , \forall \underline{U}_j, \underline{U}_t \in D_0 .$$

Suppose there exists a $\underline{U}_0 \in D_0$ such that $\| \nabla^2 G_0(\underline{U}_0)^{-1} \| \leq \beta$ and $\alpha = \beta\gamma\eta \leq \frac{1}{2}$ where $\eta \geq \| \nabla^2 G_0(\underline{U}_0)^{-1} \nabla G_0(\underline{U}_0) \|$. Set

$$t^* = (\beta\gamma)^{-1} [1 - (1 - 2\alpha)^{\frac{1}{2}}], \quad t^{**} = (\beta\gamma)^{-1}[1 + (1 - 2\alpha)^{\frac{1}{2}}] ,$$

and assume that $\{ \underline{U} \in E^{nK} \mid \|\underline{U} - \underline{U}_0\| \leq t^*\} \subset D_0$. Then the iterates (6.3.18) with $\alpha_k \equiv 1$ are well defined, remain in $\{\underline{U} \in E^{nK} \mid \|\underline{U} - \underline{U}_0\| \leq t^*\}$ and converge to a solution U^* of $\nabla G_0(\underline{U}) = \underline{0}$ which is unique in $\{\underline{U} \in E^{nK} \mid \|\underline{U} - \underline{U}_0\| < t^{**}\} \cap D_0$.

Proof

Note that $\nabla^2 G_0(\underline{U}_j) = (N_0^T G_y N_j + G_u)$ and thus $\nabla^2 G_0(\underline{U}_0)$ is given by (6.3.15).

Define $\Gamma : D_0 \subset E^{nK} \to E^{nK}$, $\Gamma\underline{U} = \underline{U} - (N_0^T G_y N_0 + G_u)^{-1} \nabla G_0(\underline{U})$; then $\nabla\Gamma(\underline{U}) = I - (N_0^T G_y N_0 + G_u)^{-1} \nabla^2 G_0(\underline{U})$, so that

$$\|\nabla\Gamma(\underline{U}_j) - \nabla\Gamma(\underline{U}_t)\| = \|\nabla^2 G_0(\underline{U}_0)^{-1}[\nabla^2 G_0(\underline{U}_j) - \nabla^2 G_0(\underline{U}_t)]\| \leq \beta\gamma \|\underline{U}_j - \underline{U}_t\|$$

$$\forall \underline{U}_j, \underline{U}_t \in D_0 ,$$

and $\nabla\Gamma(\underline{U}_0) = \underline{0}$. The result follows directly from Ortega and Rheinboldt (1970, Theorem 12.5.5). □

The rate of convergence of the simplified Gauss-Newton method is usually linear. The only exception to this is given in the following theorem.

Theorem (6.2)

The simplified Gauss-Newton iteration (6.3.18) with $\alpha_k \equiv 1$ converges to $\underline{U}^*$ Q-superlinearly (i.e. $\lim_{k\to\infty} \|\underline{U}^* - \underline{U}_{k+1}\| / \|\underline{U}^* - \underline{U}_k\| = 0$) if $N_o = N_*$ where $N_* \triangleq \partial\underline{Y}/\partial\underline{U}|_{\underline{U}=\underline{U}^*}$.

Proof

We can write

$$\Gamma(\underline{U}^*) = \underline{U}^* - (N_o^T G_y N_o + G_u)^{-1} \nabla G_o(\underline{U}^*)$$

and

$$\nabla\Gamma(\underline{U}^*) = I - (N_o^T G_y N_o + G_u)^{-1} (N_o^T G_y N_* + G_u) = \underline{0}$$

which holds if $N_o = N_*$. The result follows from Ortega and Rheinboldt (1970, Theorem 10.1.6) from which R-superlinear convergence may also be shown. □

The special case $N_o = N_*$ discussed in Theorem (6.2) occurs rather frequently in economic policy optimisation. In Section (6.2) it was shown that the simulation of an econometric model along a given path, say $\underline{U}^d$, is equivalent to solving (6.1.2) with $G_y \equiv 0$. Similarly, $G_y >> G_u$ implies $\underline{U}^* \cong \underline{U}^d$. This can be verified using (6.3.6). Since $\nabla G(\underline{U}^*) = \underline{0}$ is a necessary condition for optimality, setting $G_u = \tau G'$ where G' is positive definite and the scalar $\tau > 0$,

$$\lim_{\tau\to\infty} \underline{U}^* = \lim_{\tau\to\infty} [\underline{U}^d - (\tau G')^{-1} N_*^T G_y(\underline{g}(\underline{U}^*) - \underline{Y}^d)] = \underline{U}^d .$$

Thus, as the elements of G_u are made adequately large then $\underline{U}^* \cong \underline{U}^d$ holds. The relation $\underline{U}^* \cong \underline{U}^d$ is a well known requirement in policy optimisation since policy makers do not want significant deviations of optimal policy instrument values from $\underline{U}^d$. In view of this the initial trajectory in Algorithm (6.3.11) is usually chosen as $\underline{U}_o = \underline{U}^d$ and $\underline{Y}_o = \underline{g}(\underline{U}_o)$. Given the relative magnitudes of the weighting matrices, $G_u >> G_y$[1], $\underline{Y}_o, \underline{U}_o$ are very close to the optimum solution. Thus $N_o \cong N_*$. In such a framework Theorem (6.2) would almost hold. This explains the favourable numerical results reported by Rustem and Zarrop (1979,a) in relation to the simplified Gauss-Newton algorithm. Furthermore, numerical experiments with varying G_y have indicated that the simplified Gauss-Newton algorithm becomes inadequate as the elements of G_y are increased relative to G_u (see Rustem and Zarrop (1979,b)). Problems involving relatively high values of G_y arise when optimal policy instruments are to be computed that attain the desired endogenous values, $\underline{Y}^d$.

A further simplification to the simplified Gauss-Newton algorithm is the approximation of N_o by the dynamic multipliers of the econometric model, generated from the control origin, $\underline{Y}_o, \underline{U}_o$. The accuracy of this approximation has to be tested for each model. The elements of $N_o = \partial\underline{Y}/\partial\underline{U}|_{\underline{U}=\underline{U}_o}$ are

$$\frac{\partial y_i(\ell)}{\partial u_j(k)} \qquad 1 \le k,\ \ell \le K, \quad 1 \le i \le m, \quad 1 \le j \le n$$

where m,n,k are the number of endogenous variables, policy instruments and time periods respectively. It is assumed that

$$\frac{\partial y_i(\ell)}{\partial u_j(k)} = 0 \qquad \forall\ \ell < k$$

---

1 Assuming τ is adequately large, the elements of $G_u = \tau G'$ are much larger, in absolute value, than those of G_y. We denote this by $G_u >> G_y$; $G_y >> G_u$ denotes the converse when $|\tau|$ is adequately small (i.e. $\tau \simeq 0$).

which implies that changes in the policy instrument values can only affect current and future endogenous values. For constructing N_0 with the dynamic multipliers, the further assumption

Assumption (6.2)

$$\frac{\partial y_i(\ell)}{\partial u_j(k)} = M_{ij}(\ell - k) \quad , \quad \forall\, k \le \ell$$

has to be made. The matrix $M(\tau)$ is the dynamic multiplier at lag τ. Assumption (6.2) is only true for a time invariant linear model. However, using it avoids the numerical differentiation of the model equations since N_0 given by

$$N_0 = \begin{bmatrix} M(0) & 0 \cdots\cdots\cdots 0 \\ M(1) & M(0) \cdots\cdots 0 \\ \vdots & \vdots \qquad\quad \vdots \\ M(K-1) & M(K-2) \cdots M(0) \end{bmatrix} \qquad (6.3.19)$$

can be numerically constructed by solving the model n times rather than nK times.

Preston et al (1976) discuss the application of an algorithm using (6.2) to the Wharton Long Term Annual and Industry Forecasting Model. Rustem and Zarrop (1979,a) discuss the numerical results obtained by applying the simplified Gauss-Newton algorithm with Assumption (6.2) and (6.3.19) to the London Business School model of U.K. economy.

6.4 A QUASI-NEWTON ALGORITHM

Two basic and interconnected problems related to the optimisation of nonlinear econometric models are discussed in this section. The first is the question of evaluating the matrix N_k given by (6.3.7). The second problem is preserving the inherent lower block triangular structure of N_k in econometric models. The main reason for this structure arises from the reasonable assumption that changes in the policy instrument values affect only current and future endogenous values and not the past. The central problem, however, is to avoid the explicit evaluation of N_k by numerical differentiation.

It turns out that both these problems may be resolved within the framework of quasi-Newton algorithms. The rank-one formula due to Broyden (1965) may be adopted for updating an approximation to N_k at every iteration. The structure of N_k may be preserved using Schubert's (1970) modification of Broyden's formula.

The rank-one formula used for computing an approximation to N_{k+1} given $\underline{U}_k, \underline{U}_{k+1}$ and N_k is given by

$$N_{k+1} = N_k + \frac{(\underline{g}(\underline{U}_{k+1}) - \underline{g}(\underline{U}_k) - \alpha_k N_k \underline{d}_k^o)\underline{d}_k^{o\,T}}{\alpha_k < \underline{d}_k^o, \underline{d}_k^o >} \qquad (6.4.1)$$

Setting

$$\sigma = \frac{1}{\alpha_k < \underline{d}_k^o, \underline{d}_k^o >} \qquad (6.4.2,a)$$

$$\underline{v}_k = \underline{Y}_{k+1} - \underline{Y}_k - \alpha_k N_k \underline{d}_k^o, \qquad (6.4.2,b)$$

(6.4.1) may be written as

$$N_{k+1} = N_k + \sigma \underline{v}_k \underline{d}_k^{o\,T} . \tag{6.4.3}$$

Formula (6.4.1) is due to Broyden (1965) and has been used by Powell (1970) in his least squares approach for solving nonlinear equations. Both Broyden and Powell have considered the case when the number of equations is equal to the number of variables.

The rank-one formula (6.4.3) may also be used to compute updates to G_k^o. Thus, instead of using (6.3.10) to compute G_{k+1}^o given N_{k+1}, we can write

$$\begin{aligned} G_{k+1}^o &= N_{k+1}^T G_y N_{k+1} + G_u \\ &= (N_k + \sigma \underline{v}_k \underline{d}_k^{o\,T})^T G_y (N_k + \sigma \underline{v}_k \underline{d}_k^{o\,T}) + G_u \\ &= G_k^o + \rho[(N_k^T G_y \underline{v}_k + \mu \underline{d}_k^o)(N_k^T G_y \underline{v}_k + \mu \underline{d}_k^o)^T - N_k^T G_y \underline{v}_k \underline{v}_k^T G_y N_k] \end{aligned} \tag{6.4.4}$$

where

$$\rho = \langle \underline{v}_k , G_y \underline{v}_k \rangle^{-1} \quad \text{and} \quad \mu = \sigma\rho^{-1} . \tag{6.4.5}$$

The descent direction $\underline{d}_k^o$ given by (6.3.10) is computed by solving

$$G_k^o \underline{d}_k^o = - \nabla G(\underline{U}_k) \tag{6.4.6}$$

for $\underline{d}_k^o$. The solution to this system may be efficiently obtained by recurring the Cholesky factors of G_k^o such that

$$G_k^o = L_k D_k L_k^T \tag{6.4.7}$$

where L_k is a unit-lower triangular and D_k is a diagonal matrix. Thus, (6.4.6) can be written as

$$L_k D_k L_k^T \underline{d}_k^o = - \nabla G(\underline{U}_k)$$

from which $\underline{d}_k^o$ may be obtained by solving

$$L_k \underline{u} = - \nabla G(\underline{U}_k)$$

for $\underline{u}$, and

$$L_k^T \underline{d}_k^o = D_k^{-1} \underline{u}$$

for $\underline{d}_k^o$, using a forward and backward substitution.[1] When G_{k+1}^o is given by (6.4.4), the Cholesky factors of G_{k+1}^o may be obtained by modifying the factors of H_k^o given by (6.4.7). Methods for modifying (6.4.7) to obtain the factors of (6.4.4) consist of two successive steps

$$\bar{L}\bar{D}\bar{L}^T = L_k D_k L_k^T + \rho(N_k^T G_y \underline{v}_k + \mu \underline{d}_k^o)(N_k^T G_y \underline{v}_k + \mu \underline{d}_k^o)^T \qquad (6.4.8,a)$$

$$L_{k+1} D_{k+1} L_{k+1}^T = \bar{L}\bar{D}\bar{L}^T - N_k^T G_y \underline{v}_k \underline{v}_k^T G_y N_k \qquad (6.4.8,b)$$

Both (6.4.8,a) and (6.4.8,b) require a method to factorise the result of a rank-one modification to the given Cholesky factors. Methods for computing (6.4.8) are discussed in detail by Gill and Murray (1972) and Gill et al (1974). To summarise one such method let

$$G^* = L^* D^* L^{*T} = G + \tau \underline{z}\underline{z}^T = LDL^T + \tau \underline{z}\underline{z}^T \qquad (6.4.9)$$

where G^* and $L^* D^* L^{*T}$ denote the left side of either (6.4.8,a) or (6.4.8,b), LDL^T denotes the given Cholesky factor form of G on the right and $\tau \underline{z}\underline{z}^T$ denotes the rank - one modification. Writing (6.4.9) as $G^* = L(D + \tau \underline{w}\underline{w}^T)L^T$, where $L\underline{z} = \underline{w}$, and computing

---

1 An equivalent representation of (6.3.14,c) may be obtained by using Cholesky factors of $G_y = L_y L_y^T$ and $G_u = L_u L_u^T$, where L_y is a lower trapezoidal and L_u is a lower triangular matrix. The Gauss-Newton search direction $\underline{d}_k^o = (\underline{U} - \underline{U}_k)$ is then the solution of

$$\min_{\underline{d}_k^o} \tfrac{1}{2} \left\| \begin{bmatrix} L_y^T N_k \\ L_u^T \end{bmatrix} \underline{d}_k^o - \begin{bmatrix} \underline{Y}^d - \underline{Y}_k \\ \underline{U}^d - \underline{U}_k \end{bmatrix} \right\|^2 .$$

$$\tilde{L}\tilde{D}\tilde{L}^T = D + \underline{w}\,\underline{w}^T$$

by the method described in Gill and Murray (1972) we have $L^* = L\tilde{L}$ and $D^* = \tilde{D}$. Other approaches to computing and updating Cholesky factors of G^o_{k+1} are discussed by Goldfarb (1976).

Since $G_k(\underline{U})$ (6.3.14,c) is the predicted value of $G(\underline{U})$ given the linearisation

$$\underline{g}(\underline{U}) \cong g(\underline{U}_k) + N_k(\underline{U} - \underline{U}_k) \tag{6.4.10}$$

a comparison between $G_k(\underline{U})$ and $G(\underline{U})$ can be used for testing the adequacy of the linear approximation (6.4.10) over the distance $||\underline{U} - \underline{U}_k||$. The test

$$G(\underline{U}_k) - G(\underline{U}_{k+1}) \geq \gamma[G(\underline{U}_k) - G_k(\underline{U}_{k+1})] \quad , \tag{6.4.11}$$

with $\gamma \in (0,1)$, measuring the ratio of the actual decrease in the objective function to the predicted decrease, is suggested by Powell (1970). Thus, if $\underline{U}_{k+1}$ satisfies (6.4.11) the approximation (6.4.10) is valid for the distance $\alpha_k\,||\underline{d}^o_k||$ along $\underline{d}^o_k$. Otherwise α_k has to be reduced until (6.4.11) is satisfied. However, since

$$\begin{aligned} G(\underline{U}_k) - G_k(\underline{U}_{k+1}) &= \tfrac{1}{2}[||\underline{Y}_k - \underline{Y}^d||^2_{G_y} + ||\underline{U}_k - \underline{U}^d||^2_{G_u} \\ &\quad - ||\alpha_k N_k \underline{d}^o_k + \underline{Y}_k - \underline{Y}^d||^2_{G_y} - ||\alpha_k \underline{d}^o_k + \underline{U}_k - \underline{U}^d||^2_{G_u}] \\ &= -\tfrac{1}{2}\alpha_k^2 ||N_k \underline{d}^o_k||^2_{G_y} - \alpha_k < N_k\underline{d}^o_k, G_y(\underline{Y}_k - \underline{Y}^d) > \\ &\quad - \tfrac{1}{2}\alpha_k^2 ||\underline{d}^o_k||^2_{G_u} - \alpha_k < \underline{d}^o_k, G_u(\underline{U} - \underline{U}^d_k) > \\ &= -\tfrac{1}{2}\alpha_k^2 < \underline{d}^o_k, (N_k^T G_y N_k + G_u)\underline{d}^o_k > - \alpha_k < \underline{d}^o_k, \nabla G(\underline{U}_k) > \end{aligned} \tag{6.4.12}$$

if $-\alpha_k < \underline{d}^o_k, \nabla G(\underline{U}_k) > \; \geq -\tfrac{1}{2}\alpha_k^2 < \underline{d}^o_k, (N_k^T G_y N_k + G_u)\underline{d}^o_k >$, the criterion

$$G(\underline{U}_k) - G(\underline{U}_{k+1}) \geq -\gamma\alpha_k < \nabla G(\underline{U}_k), \underline{d}_k^o > \qquad (6.4.13)$$

used in (6.3.12) also serves as the required test for the validity of the linear approximation. Furthermore, for $\underline{d}_k^o$ given by (6.3.10)

$$G(\underline{U}_k) - G_k(\underline{U}_{k+1}) = -\frac{\alpha_k^2}{2} < \nabla G(\underline{U}_k), H_k^o \nabla G(\underline{U}_k) > \qquad (6.4.14)$$

$$+ \quad \alpha_k < \nabla G(\underline{U}_k), H_k^o \nabla G(U_k) >$$

$$> \quad 0$$

for $2 > \alpha_k > 0$ and $\nabla G(\underline{U}_k) \neq \underline{0}$. It follows therefore that α_k satisfying (6.4.11) or (6.4.13) reduces the objective function so that

$$G(\underline{U}_k) - G(\underline{U}_{k+1}) > 0 .$$

The algorithm below aims to compute $\underline{d}_k^o$ and α_k such that the sequence $\underline{U}_k$, $k = 0,1,\ldots$ corresponds to successive reductions in the objective function.

In order to capture a uniformly balanced information in all directions about the gradients ∇g_i, N_k needs to be updated along directions which are uniformly linearly independent. A definition of uniformly linearly independent vectors is given in Ortega and Rheinboldt (1970) and a method for ensuring that N_k is updated along uniformly independent directions is given by Powell (1970, a,b). Thus using only $\underline{d}_k^o$ in (6.4.3) as the direction along which N_k is updated is not sufficient. However, an alternative to imposing this condition on the directions is re-evaluating N_k by numerical differentiation after each $k_o \geq nK$ iterations. The algorithm below employs the latter alternative. The initial estimate of the optimal policy vector $\underline{U}_o$, supplied to the algorithm, is used to compute $\underline{Y}_o = \underline{g}(\underline{U}_o)$ and N_o. The initial approximation N_o may be obtained by using the dynamic multipliers of the model (Rustem and Zarrop (1979,a)). The algorithm summarised below also requires the numerical evaluation of N to verify convergence to the optimum solution. The dynamic multipliers may also be used at this stage as an approximation to numerical derivatives. The use of the dynamic multipliers for this purpose is less desirable and can only be justified in the absence of a better approximation to the derivatives. In such cases numerical experience indicates that the periodic re-evaluation of N_k at intervals of $k_o \geq nK$ iterations as discussed above, have to be abandoned rather than replacing N_k by the dynamic multipliers. Experiments have shown that N_k should be replaced by the dynamic multipliers only when a descent direction cannot be obtained using the current N_k.

<u>Algorithm</u> (6.4.15)

<u>Step 0:</u> Given $\underline{Y}^d$, $\underline{U}^d$, G_y, G_u, $\underline{U}_o$, compute $\underline{Y}_o = \underline{g}(\underline{U}_o)$; set q_{max} to the largest element of G_y, $\phi = q_{max}\,\varepsilon_o$ where $\varepsilon_o \in (0,\ 5.0\text{x}10^{-4}]$ is the accuracy of the model solution program, $\nu\in(0,\ 10^{-4}]$, $\mu\in(0,\ 1)$, $\delta\in(0,\ 10^{-5}]$, $\eta\in(0,\ 1)$, $k = 0$.

<u>Step 1:</u> Compute N_k by numerical differentiation.

<u>Step 2:</u> Compute $\nabla G(\underline{U}_k)$.

<u>Step 3:</u> Optimality check: if $\| \nabla G(\underline{U}_k) \| \leq \phi$

or if $k > o$ $\| \underline{U}_{k-1} - \underline{U}_k \|_2 \leq \nu$ and $G(\underline{U}_{k-1}) - G(\underline{U}_k) \leq \phi$ (6.4.16)

compute N_k, $\nabla G(\underline{U}_k)$ in Steps 1 - 3

and if (6.4.16) is still satisfied, stop.

<u>Step 4:</u> Compute $\underline{d}_k^o$.

<u>Step 5:</u> Compute the step size α_k: Set α_k to the first member of the sequence $\{(\eta)^j;\ j=0,1,\ \ldots\}$ such that (6.4.11) or (6.4.13) is satisfied. If

$$\alpha_k < \delta \tag{6.4.17}$$

compute steps 1 - 5 and if (6.4.17) is still satisfied, stop.

Step 6: Set

$$\underline{U}_{k+1} = \underline{U}_k + \alpha_k \underline{d}_k^o . \tag{6.4.18}$$

Step 7: If $\|\underline{Y}_{k+1} - \underline{Y}_k\|_2 \leq \varepsilon_o$ go to Step 8, otherwise update N_k using (6.4.3) to obtain N_{k+1} and update G_k^o using (6.4.4) to obtain G_{k+1}^o.

Step 8: Set $k = k+1$, to to Step 2.

In the numerical experiments reported in Rustem and Zarrop (1979,b) and Karakitos, Rustem, Zarrop (1979) the above algorithm was used with $\phi = \nu = 10^{-4}$, $\eta = \mu = .1$, $\delta = 10^{-5}$, $\varepsilon_o = 5 \times 10^{-4}$. The restriction $\|\underline{Y}_{k+1} - \underline{Y}_k\|_2 > \varepsilon_o$ is imposed to avoid updating N_k when $\underline{Y}_{k+1} - \underline{Y}_k$ is dominated by the errors in solving the model for $\underline{Y}_{k+1}$ and $\underline{Y}_k$.

The algorithm uses N_k computed by (6.4.3) when it can be used to obtain a descent direction. Thus, if (6.4.13) is not satisfied for $\alpha_k > \delta$ with the approximation given by (6.4.3), N_k is re-evaluated by numerical differentiation. Furthermore, if N_k is re-evaluated periodically at every $k_o \geq nK$ iterations, the convergence of the sequences $\{\underline{U}_k\}$ and $\{\underline{d}_k^o\}$, i.e.

$$\lim_{k\to\infty} \underline{U}_k = \underline{U}^* \qquad \text{and} \qquad \lim_{k\to\infty} \underline{d}_k^o = \underline{0}$$

follow from the convergence of the damped Gauss-Newton algorithm (Ortega and Rheinboldt (1970, Theorem 14.4.4)). In this case Powell's results (see Powell (1970,a , Theorems 3 and 4) imply that N_k

defined by (6.4.3) with $\alpha_k \equiv 1$ is bounded and converges to N_* (see also Moré and Trangenstein (1976, Theorems 5.6 and 5.7).

In order to preserve the block lower diagonal structure of N_k only those elements that are not constants may be updated. The updating of these elements is done so as to account for those residual changes in the nonlinear equations which cannot be accounted by the fixed (e.g. zero) elements of N_k. This is basically the modification to Broyden's (1965) method, due to Schubert (1970). The updating is simply accomplished as follows:

Definition (6.1)

We define the row vector $\underline{n}_k^i$ as the ith row of the matrix N_k.

Definition (6.2)

For the ith row of N_k define a column vector $\underline{d}_k^i$ derived from $\underline{d}_k^o$ in (6.3.10) by setting equal to zero those elements in $\underline{d}_k^o$ which correspond to constant (e.g. zero) values in $\underline{n}_k^i$.

The resulting updating formula for those elements of N_k which are not constants is given for each row of N_{k+1} since $\underline{d}_k^i$ depends on the row being calculated. Hence

$$\underline{n}_{k+1}^i = \underline{n}_k^i + \frac{(Y_{k+1}^i - Y_k^i - {}_k\underline{n}_k^i \underline{d}_k^o)\, \underline{d}_k^{i\,T}}{\alpha_k < \underline{d}_k^i, \underline{d}_k^i >} \tag{6.4.19}$$

may be used to compute the ith row of N_{k+1}, $i = 1,2,\ldots, mK$. The superscript on Y in (6.4.19) denotes the ith element of the vector $\underline{Y}$. The convergence properties of Schubert's modification have been discussed by Broyden (1971) and Marvill (1978).

6.5 CONCLUDING REMARKS

The static optimisation framework adopted in this study inevitably entails an open loop approach to the optimal control problem with nonlinear models (6.1.2). By this we mean that the optimal values $\underline{U}^*$ are computed, ahead of time, for the period $1 \le k \le K$ (see (5.2.2) - (5.2.3)). A closed-loop or feedback control of the dynamic system $\underline{F}(\underline{Y}, \underline{U}) = \underline{0}$ would also require the possibility of $\underline{u}^*(k)$, the optimal value of $\underline{U}^*$ at time k, to depend on the evolution of the dynamic system up to time k. This would be especially important in stochastic systems. To account for such effects, Athans et al (1976) have described a sequential procedure for updating the optimisation problem (6.1.2) with incoming information about exogenous assumptions. Thus (6.1.2) is solved a number of times, for gradually shorter overall periods K. Each time the optimisation origin is moved forward a few periods and the exogenous assumptions are updated.

Imposing linear inequality constraints (e.g. bounds) on $\underline{Y}$ and $\underline{U}$ may be avoided by altering the objective function (see Chapter 5). However, if the number of constraints to be imposed is large, then adopting the algorithms in Sections (6.3) - (6.4) to inequality constraints is preferable. For linear constraints involving $\underline{U}$ only, this can be suitably accomplished by adopting an active set strategy (see Gill and Murray (1974)). A graduate student at Imperial College, Vallet (1977), has studied this problem in connection with a linear econometric model.

APPENDIX : NUMERICAL RESULTS

The behaviour of the optimal value of the objective function for increasing values of G_u is related to $\underline{U}^*$. From the discussion in Section (6.3), as $\tau \to \infty$, $\underline{U}^* \to \underline{U}^d$ and thus for suitable large values of G_u, the policy instruments attain their desired values. A similar result for G_y would indicate that achieving $\underline{Y}^* \cong \underline{Y}^d$ might be possible for large values of G_y. To show that if the ith diagonal element of G_y, $[G_y]_{ii} \to \infty$ then $Y^{i^*} \to Y_i^d$, consider (6.3.6) at $\underline{U}^*$. Since $\nabla G(\underline{U}^*) = \underline{0}$ is a necessary condition for optimality, dividing it through by $[G_y]_{ii}$ yields the required result as the only solution of (6.3.6) is given by $Y^{i^*} = Y_i^d$, provided the ith column of N_* is not a zero vector. Policy optimisation exercises have generally been confined to cases in which the elements of G_y are at the same order or smaller than those of G_u (see, e.g. Holly et al (1979), Klein (1979)). This is due to the popular belief that if $G_y >> G_u$ [1] the optimal policy instruments may be driven far away from their desired values. Such a departure may result in a high optimal objective function value. However, this may be acceptable if attaining $Y^{i^*} = Y_i^d$ is important. In an attempt to illustrate this point, an example was set up using a quarterly nonlinear econometric model of the West German economy, discussed in Karakitsos, Rustem and Zarrop (1979). The model is dynamic and has 29 equations. The optimisation study was done for 20 quarters (i.e. 5 years). The objective function is given in Table 1. In this exercise the diagonal weight on $y_3(k)$, $1 \le k \le 20$ was changed from 1 to 10^{10}. The desired values and resulting optimal trajectories of y_3 are plotted in

---

[1] See footnote on page 272.

Table 1:

The Specification of the Objective Function
for the period 1973(1) - 1977(4)

Policy Instruments	Weighted Endogenous Variables
u_1 : monetary base	y_1 : level of unemployment
u_2 : discount rate	y_2 : rate of inflation
	y_3 : change in money stock
	y_4 : rate of growth of the economy
	y_5 : rate of change in the money stock

Desired Values

Policy instruments: (the historical paths were chosen)

$u_1^d(k)$; $1 \le k \le 20$

90.9, 92.6, 93, 94.7, 96, 97.8, 99.3, 100.7, 103, 104.8, 107.4, 110.6, 112.2, 114.8, 119.9, 112.3, 124.8, 128.4, 131.8 .

$u_2^d(k)$; $1 \le k \le 20$ (in %)

5, 7, 7, 7, 7, 7, 7, 6, 5, 4.5, 3.5, 3.5, 3.5, 3.5, 3.5, 3.5, 3.5, 3.5, 3.5, 3 .

Endogenous Values :

$y_1^d(k)$; $1 \le k \le 4$

250000, 260000, 270000, 280000

$5 \le k \le 20$; constant at 300,000 for all k

$y_2^d(k)$; $1 \le k \le 20$

1.0024, for all k .

$y_3^d(k)$; $1 \le k \le 20$

1.0171 for all k

Table 1 (continued)

$y_4(k)$; $1 \le k \le 20$

1.0008, .98706, 1.008, 1.0052, 1.0242, .99625, .99904, .98994, 1.0016, 1.0196, 1.0152, 1.0147, 1.0079, 1.0016, 1.0061, 1.0027, 1.0066, 1.0054, 1.0053, 1.0089.

$y_5(k)$; $1 \le k \le 20$

0 for all k.

Weights: only the diagonal weights are specified: the off diagonal elements are set to zero

Q_u : Diagonal weights: for all k, $1 \le k \le 20$

u_1: 2×10^{-3}

u_2: 4×10^{-3}

Q_y : Diagonal weights: for all k, $1 \le k \le 20$

y_1: 10^{-4}

y_2: 5×10^{3}

y_3: varied between 1 and 10^{10} for different runs.

y_4: 10^{16}

y_5: 10^{10}.

Figure 1. The behaviour of the algorithm is summarised in Table 2.

Numerical results concerning the application of the simplified Gauss-Newton algorithm, (6.3.15) - (6.3.18), to the London Business School model of the U.K. economy was reported in Rustem and Zarrop (1979,a). In Rustem and Zarrop (1979,b) the application of algorithm (6.4.15) to an econometric model of the Netherlands is discussed. Numerical results from this application have indicated that for $G_u >> G_y$ the simplified Gauss-Newton algorithm performs reasonably well compared to (6.4.15). This supports the argument in Section (6.3) related to the simplified Gauss-Newton algorithm. However, when, G_u is no longer large enough numerical evidence has shown that algorithm (6.4.15) is more appropriate.

Table 2: The Performance of Algorithm (6.4.15)

Diagonal weight of G_y on y_3	1	10^2	10^3	10^4	10^5	10^6	10^7	10^8	10^9	10^{10}
Number of evaluations of N_k by approximate differentiation (i.e. using the dynamic multipliers)	2	2	3	2	2	6	2	2	2	2
Number of times the Quasi-Newton formula (4.3) used to update N_k	9	27	28	8	22	13	9	6	6	6
Number of iterations taken to converge	10	28	29	9	30	14	14	13	13	13
Optimal value of the objective function	$.431 \times 10^{11}$	$.300 \times 10^{11}$	$.840 \times 10^9$	$.952 \times 10^7$	$.204 \times 10^4$	$.152 \times 10^3$	$.978 \times 10^2$	$.978 \times 10^2$	$.978 \times 10^2$	$.978 \times 10^2$
CDC Cyber 172 time in seconds taken to solve the problem	30.515	73.563	79.568	26.623	80.549	50.294	31.996	28.417	28.542	28.105

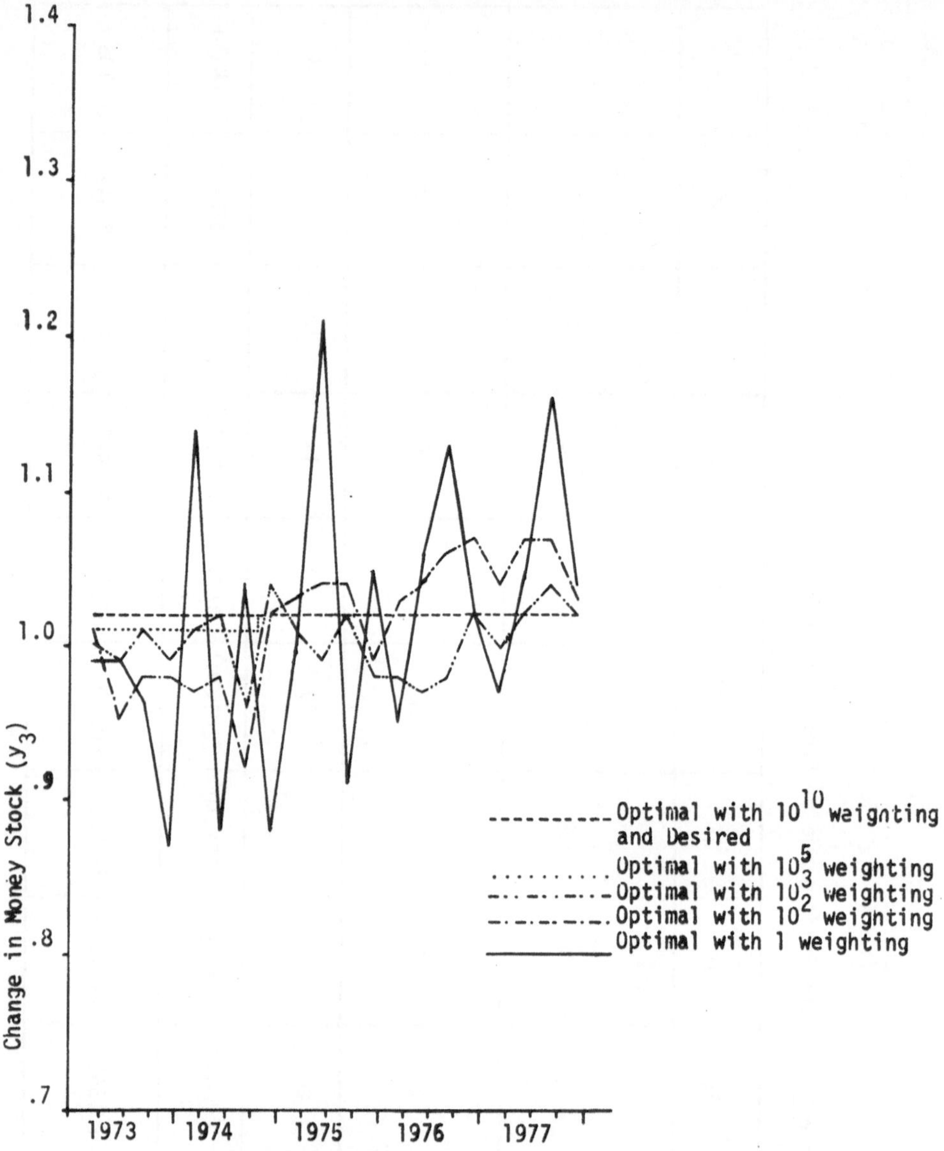

Figure 1: Plot of Desired and Optimal Values of y_3 for varying weights.

CHAPTER 7

SUGGESTIONS FOR FURTHER RESEARCH

In this work the application of projection methods to problems in constrained optimisation have been discussed. The techniques discussed in Chapters 1,2,3,4 and 6 are for deterministic optimisation problems in which both the objective function and the constraint evaluations are exactly measurable. However, optimal decisions are mostly taken in an uncertain environment (see, e.g. Theil (1964), Kornai (1975)). Such problems can be characterised in terms of stochastic optimisation. This leads to a general class of optimisation problems in which there exist errors in the measurement of the objective function and/or the constraint values (Kushner and Clark(1978)).The extensions of some deterministic algorithms to stochastic problems are discussed by Polyak(1978) and by Kushner and Clark (1978) who also establish convergence properties. However, because of the heavy computational requirements involved, the application of such algorithms have hitherto been confined to problems with very few variables (see, e.g. Farah (1977) for 2 - 4 variable examples). With the advent of array processing technology in computers (see, Parkinson (1978)), this computational restriction is likely to be reduced and thus the extension of deterministic methods to stochastic optimisation problems with *relatively* large number of variables will be feasible. Hence one field of further investigation is the extension of the methods in Chapters 1,2,3,4 and 6 to stochastic problems on computers with array processors.

The algorithm in Chapter 5 is useful for the iterative respecification of the weighting (i.e. Hessian) matrix of the quadratic objective function. The extensions of this algorithm can be considered for characterising bargaining problems in which all agents simultaneously minimise their objective functions. Each bargaining agent is allowed his own quadratic objective function. The desired values are chosen. As the bargaining proceeds, each agent's preferance ranking changes, in attaining his desired values, given his opponents' revealed preferances. Thus, each bargaining agent, in turn, alters the Hessian of his objective function to reflect his changing preferances. After making such an alteration, the agent recomputes the optimal variables and submits them to his opponents as the values he would prefer to see. A statement of this extension of the respecification method in Chapter 5 to bargaining problems is discussed in Rustem and Velupillai (1979).

Finally, the optimisation algorithms in this work can be extended to provide solutions to multiple objective optimisation problems. These can be formulated as the joint minimisation of a number of objective functions. An attempt in this direction is discussed in Rustem and Zarrop (1980) in which Algorithm (6.4.15) is extended to solve Nash equilibrium problems with two quadratic objective functions and a nonlinear model. The numerical results obtained from the application of this algorithm to a nonlinear econometric model of the Netherlands are also discussed in Rustem and Zarrop (1980).

REFERENCES FOR CHAPTER I

Abadie, J. and Carpentier, J. (1969): "Generalisation of the Wolfe Reduced Gradient Method to the Case of Nonlinear Constraints", in: Optimisation, R. Fletcher (ed.), Academic Press, London and New York.

Bartels, R.H., Golub, G.H. and Saunders, M.A. (1970): "Numerical Techniques in Mathematical Programming", in: Nonlinear Programming, J.B. Rosen, O.L. Mangasarian and K. Ritter (eds.), Academic Press, New York.

Beale, E.M.L. (1967): "Numerical Methods" in: Nonlinear Programming, J. Abadie (ed.), North-Holland, Amsterdam.

Bertsekas, D.P. (1976): "On the Goldstein-Levitin Polyak Gradient Projection Method", I E E E Trans. AC, V.21, pp. 174-183.

Biggs, M.C. (1974): The Development of a Class of Constrained Optimisation Algorithms and their Application to the Problem of Electric Power Scheduling, Ph.D. Thesis, University of London.

Biggs, M.C. (1975): "Constrained Minimisation Using Recursive Quadratic Programming: Some Alternative Subproblem Formulations", in: Towards Global Optimisation, L.C.W Dixon and G.P. Szego (eds.) North-Holland, Amsterdam.

Buckley, A.G. (1975): "An Alternative Implementation of Goldfarb's Minimization Algorithm", Math. Prog., V.8, pp. 207-231.

Davidon , W. (1959): "Variable Metric Methods for Minimization" , A E C Res. and Devel. Report, ANL-5990, Argonne National Laboratory, Lemont, Illionois.

Davis, M.H.A. (1977): Linear Estimation and Stochastic Control, Chapman and Hall, London,

Dennis, J.B. (1959): Mathematical Programming and Electrical Networks, Wiley, New York.

Fiacco, A.V. and McCormick, G.P. (1968): Nonlinear Programming: Sequential Unconstrained Minimization Techniques, J. Wiley, New York.

Fletcher, R. (1972): "An Algorithm for Solving Linearly Constrained Optimization Problems" , Math. Prog., V.2, pp. 133-165.

Ganzhela, I.F. (1970): "An Algorithm of Descent with Constraints" U S S R Comp. Math. and Math. Phys. , V.10, pp. 191-206.

Garcia-Palomares, U.M. and Mangasarian, O.L. (1974): "Superlinearly Convergent Quasi-Newton Algorithms for Nonlinearly Constrained Optimization Problems" , Comp. Sciences Tech.: Rep. No. 195, University of Wisconsin, Madison.

Gill, P.E., Golub, G.H., Murray, W. and Saunders, M.A. (1974): "Methods for Modifying Matrix Factorisations", Maths. of Comp., V.28, pp. 505-535.

Gill, P.E. and Murray, W. (1972): "Two Methods for the Solution of Linearly Constrained and Unconstrained Optimisation Problems" , N P L Report NAC 25 , National Physical Laboratory, Teddington.

Gill, P.E. and Murray, W. (1973): "Quasi-Newton Methods for Linearly Constrained Optimisation" , N P L Report NAC 32, National Physical Laboratory, Teddington.

Gill, P.E. and Murray, W. (1974,a): "Newton-type Methods for Linearly Constrained Optimisation" , in: Numerical Methods for Constrained Optimization, P.E. Gill and W. Murray (eds.), Academic Press, London.

Gill, P.E. and Murray, W. (1974,b): "Quasi-Newton Methods for Linearly Constrained Optimization", in: Numerical Methods for Constrained Optimization, P.E. Gill and W. Murray (eds.), Academic Press, London.

Goldfarb, D. (1969): "Extension of Davidon 's Variable Metric Method to Maximization Under Linear Inequality and Equality Constraints", S I A M J. Appl. Math., V.17, 739-764.

Goldfarb, D. (1975): "Matrix Factorisations in Optimisation of Nonlinear Functions Subject to Linear Constraints, Math. Prog., V.10, pp. 1-31.

Goldstein, A.A. (1964): "Convex Programming in Hilbert Space" , Bull.Amer. Math. Soc., V.70, pp. 709-710.

Han, S.-P. (1976): "Superlinearly Convergen Variable Metric Algorithms for General Nonlinear Programming Problems", Math. Prog., V.11, pp. 263-282.

Han, S.-P. (1977,a): "A Globally Convergent Method for Nonlinear Programming" , J O T A , V.22, pp. 297-309.

Han, S.-P. (1977,b): "Dual Variable-Metric Algorithms for Constrained Optimization", S I A M J. Control & Opt., V.15, pp. 546-565.

Lemke, C.E. (1961): "The constrained Gradient Method of Linear Programming", S I A M J., V.9, pp. 1-17.

Lemke, C.E. (1968): "On Complimentary Pivot Theory", in: Mathematics of Decision Sciences, Part I , G.B. Dantzig and A.F. Veinott (eds.), American Mathematical Society.

Levitin, E.S. and Polyak, B.T. (1966): "Constrained Minimization Methods", U S S R Comp. Math. and Math. Phys., V.6, pp. 1-50.

Luenberger, D.G. (1969): Optimization by Vector Space Methods, J. Wiley, New York.

Luenberger, D.G. (1973): Introduction to Linear and Nonlinear Programming, Addison Wesley, Massachusetts.

Luenberger, D.G. (1974): "A Combined Penalty Function and Gradient Projection Method for Nonlinear Programming", J O T A, V.14, pp. 477-495.

McCormick, G.P. (1970,a): "The Variable Reduction Method for Nonlinear Programming" , Management Sci., V.17, pp. 146-160.

McCormick, G.P. (1970,b): "A Second Order Method for the Linearly Constrained Nonlinear Programming Problem", in: Nonlinear Programming , J.B. Rosen, O.L. Mangasarian and K.R. Ritter (eds.), Academic Press, New York.

McCormick, G.P. and Tapia, R.A. (1972): "The Gradient Projection Method Under Mild Differentiability Conditions", S I A M J. , Control, V.10, pp. 93-98.

Mikhlin, S.G. (1964): Variational Methods in Mathematical Physics, Pergamon Press, Oxford.

Murray, W. (1969): "Constrained Optimization", N P L Report Ma 79, National Physical Laboratory, Teddington.

Murtagh, B.A. and Sargent, R.W.H. (1969): "A Constrained Minimization Method with Quadratic Convergence", in: Optimisation, R. Fletcher (ed.), Academic Press, London and New York.

Ortega, J.M. and W.C. Rheinboldt (1970): Iterative Solution of Nonlinear Equations in Several Variables, Academic Press, London and New York.

Polak, E. (1971): Computational Methods in Optimization, Academic Press, London and New York.

Powell, M.J.D. (1974,a): "Unconstrained Minimization and Extensions for Constraints" , in: Mathematical Programming in Theory and Practice, P.L. Hammer and G. Zoutendijk (eds.), North-Holland, Amsterdam.

Powell, M.J.D. (1974,b): "Introduction to Constrained Optimization" , in: Numerical Methods for Constrained Optimization, P.E. Gill and W. Murray (eds.), Academic Press, London and New York.

Powell, M.J.D. (1976): "Algorithms for Nonlinear Constraints that Use Lagrangian Functions" , presented at the Ninth International Symposium on Mathematical Programming, Budapest.

Powell, M.J.D. (1977,a): "A Fast Algorithm for Nonlinearly Constrained Optimization Calculations", presented at the 1977 Dundee Conference on Numerical Analysis.

Powell, M.J.D. (1977,b): "The Convergence of Variable Metric Methods for Nonlinearly Constrained Optimization Calculations", Argonne National Laboratory, Appl. Math. Div. Tech. Memorandum No.315, Argonne, Illionois.

Robinson, S.M. (1972): "A Qadratically Convergent Algorithm for General Nonlinear Programming Problems", Math. Prog., V.3, pp. 145-156.

Robinson, S.M. (1974): "Perturbed Kuhn-Tucher Points and Rates of Convergence for a Class of Nonlinear Programming Algorithms", Math. Prog., V.7, pp. 1-16.

Rosen, J.B. (1960): "The Gradient Projection Method for Nonlinear Programming. Part I: Linear Constraints", S I A M J. Appl. Math., V.8, pp. 181-217.

Rosen, J.B. (1961): "The Gradient Projection Method for Nonlinear Programming. Part II: Nonlinear Constraints", S I A M J. Appl. Math., V.9, pp. 514-532.

Rosen, J.B. and Kreuser J. (1972): "A Gradient Projection Algorithm for Non-linear Constraints", in: Numerical Methods for Nonlinear Optimization, F.A. Lootsma (ed.), Academic Press, London and New York.

Rustem, B. and Velupillai, K. (1978): "On Detecting Time Varying Structures", Computing and Control Department Research Report 78/13, Imperial College, London.

Rustem, B. and Vellupillai, K. (1979): "On the Definition and Detection of Structural Change", in: Stochastic Control Theory and Stochastic Differential Systems, M. Kohlmann and W. Vogel (eds.), Springer-Verlag, Berlin.

Sargent, R.W.H. and Murtagh, B.A. (1973): "Projection Methods for Nonlinear Programming", Math. Prog., V.4, pp. 245-268.

Sargent, R.W.H. (1974): "Reduced Gradient and Projection Methods for Nonlinear Programming", in: Numerical Methods for Constrained Optimisation, P.E. Gill and W. Murray (eds.), Academic Press, London and New York.

Wilde, D.J. and Beightler, C.S. (1967): Foundation of Optimisation, Prentice-Hall Inc., Englewood Cliffs, N.J.

Wilson, R.B. (1963): A Simplicial Algorithm for Concave Programming, Ph.D. Diss. Grad. School of Business Administration, Harward University, Boston.

Wolfe, P. (1967): "Methods for Nonlinear Constraints", in: Nonlinear Programming, J. Abadie (ed.), North-Holland Publishing Co., Amsterdam.

Zoutendijk, G. (1960): "Methods of Feasible Directions", Elsevier, Amsterdam.

Zoutendijk, G. (1966): "Nonlinear Programming: A Numerical Survey", SIAM J. Control, V.4, pp. 194-210.

REFERENCES FOR CHAPTER 2

Agmon, S. (1954): "The Relaxation Method for Linear Inequalities, Canadian J. of Mathematics, V. 6, pp. 382 - 392.

Avis, D. and Chvatal, V. (1978): "Notes on Bland's Pivoting Rule", Mathematical Programming Study, V. 8, pp. 24 - 34.

Bland, R. G. (1977): "New Finite Pivoting Rules for the Simplex Method", Maths. of Operations Research, V.2, pp. 103 - 107.

Dantzig, G. (1963): Linear Programming and Extensions, Princeton University Press, Princeton, N. J.

Fletcher, R. (1970): "The Calculation of Feasible Points for Linearly Constrained Optimization Problems", A E R E Harwell Report, R. 6354.

Gill, P. E. and Murray, W. (1974): "Newton-type Methods for Linearly Constrained Optimization", in: Numerical Methods for Constrained Optimization, P. E. Gill and W. Murray (eds.) Academic Press, London.

Goldfarb, D. and Reid, J. K. (1975): "A Practicable Steepest-edge Simplex Algorithm", A E R E Harwell Report, CSS 19.

Herman, G. T. (1975): "A Relaxation Method for Reconstructing Objects from Noisy X-Rays", Math. Prog., V. 8, pp. 1 - 19.

Mangasarian, O. L. (1969): Nonlinear Programming, McGraw-Hill, New York.

Motzkin, T. S. and Schoenberg, I. J. (1954): "The Relaxation Method for Linear Inequälities", Canadian J. of Mathematics, V. 6, pp. 393 - 404.

Rockafeller, R. T. (1972): Convex Analysis, Princeton University Press, Princeton, N. J.

Rosen, J. B. (1960): "The Gradient Projection Method for Nonlinear Programming. Part I. Linear Constraints _S I A M J._, V. 8, pp. 181 - 217.

Zoutendjik, G. (1970): "Nonlinear Programming, Computational Methods", in: _Integer and Nonlinear Programming_, J. Abadie (ed.), North-Holland, Amsterdam.

REFERENCES FOR CHAPTER 3

Bartels, R. H., Golub, G. H., Saunders, M. A. (1970): "Numerical Techniques in Mathematical Programming" , in: Nonlinear Programming , J. B. Rosen, O. L. Mangasarian and K. Ritter (eds.), Academic Press, New York.

Beale, E. M. L. (1967): "Numerical Methods", in: Nonlinear Programming , J. Abadie (ed.), North-Holland, Amsterdam.

Businger, P. and Golub, G. H. (1965): "Linear Least Squares Solutions by Household Transformations" , Numerische Mathematik, V.7, pp. 269-276.

Cannon, M. D., Cullum, C. D., Polak, E. (1970): Theory of Optimal Control and Mathematical Programming , McGraw-Hill, New York.

Fiacco, A. V. and McCormick, G. P. (1968): Nonlinear Programming: Sequential Unconstrained Minimization Techniques , John wiley, New York.

Fletcher, R. (1969): "A Technique for Orthogonalisation" , J. Inst. Maths Applics , V.5, pp. 162-166.

Fletcher, R. (1971): "A General Quadratic Programming Algorithm" , J. Inst. Maths Applics , V.7, pp. 76-91.

Garcia-Palomares, U. M., and Mangasarian, O. L. (1974): "Superlinearly Convergent Quasi Newton Algorithms for Nonlinearly Constrained Optimization Problems" , Comp. Sciences Tech. Rep. No. 195 , University of Wisconsin, Madison.

Gill, P. E., Golub, G. H., Murray, W., Sunders, M. A. (1974): "Methods for Modifying Matrix Factorisations" , Maths. of Comp. , V.28, pp. 505-535.

Gill, P. E. and Murray, W. (1974): "Newton-Type Methods for Linearly Constrained Optimization" , in: Numerical Methods for Constrained Optimization, P.E. Gill and W. Murray (eds.), Academic Press, London.

Gill, P.E. and Murray, W. (1977): "The Computation of Lagrange Multiplier Estimates for Constrained Minimization", NPL Report NAC 77, Teddington.

Gill, P. E. and Murray, W. (1978): "Numerically Stable Methods for Quadratic Programming", Math. Prog., V.14, pp. 349-372.

Goldfarb, D. (1971): "Extension of Newton's Method and Simplex Methods for Solving Quadratic Programs", in: Numerical Methods for Nonlinear Optimization, F.A. Lootsma (ed.), Academic Press, London.

Goldfarb, D. (1975): "Matrix Factorizations in Optimization of Nonlinear Functions Subject to Linear Constraints", Math. Prog., V.10, pp.1-31.

Goldfarb, D. (1976): "Factorized Variable Metric Methods for Unconstrained Optimization, Math. of Comp., V.30, pp.796-811.

Golub, G. H. and Saunders, M. A. (1970): "Linear Least Squares and Quadratic Programming", in: Integer and Nonlinear Programming, J. Abadie (ed.), North-Holland, Amsterdam.

Han, S.-P. (1976): "Superlinearly Convergent Variable Metric Algorithms for General Nonlinear Programming Problems", Math. Prog., V.11, pp. 263-282.

Luenberger, D. G. (1969): Optimization by Vector Space Methods, John Wiley, New York.

Luenberger, D. G. (1973): Introduction to Linear and Nonlinear Programming, Addison-Wesley, Massachusetts.

Mazzoleni, P. (1975): "A Particular Convex Programming Algorithm", NOC Technical Report 66, The Hatfield Polytechnic.

Polak, E. (1971): Computational Methods in Optimization, Academic Press, New York.

Powell, M. J. D. (1977): "The Convergence of Variable Metric Methods for Nonlinear Constrained Optimization Calculations", Argonne National Lab., Appl. Math. Div. Tech. Memo. No. 315, Argonne, Illinois.

Theil, H. and van de Panne, C. (1960): "Quadratic Programming as an Extension of Conventional Quadratic Maximisation", MAN Management Sci., V.7, pp. 1-20.

van de Panne, C. and Whinston, A. (1969): "The Symmetric Formulation of the Simplex Method for Quadratic Programming", Econometrica, V. 37, pp. 507-527.

Wilson, R. B. (1963): A Simplicial Algorithm for Concave Programming, Ph.D. Diss. Grad. School of Business Admin. Harvard University, Boston.

Zoutendijk, G. (1970): "Nonlinear Programming, Computational Methods", in: Integer and Nonlinear Programming, J. Abadie (ed.), North-Holland, Amsterdam.

REFERENCES FOR CHAPTER 4

Armijo, L. (1966): "Minimization of Functions Having Continuous Partial Derivatives" , Pacific J. Math., V. 16, pp. 1 - 3.

Bertsekas, D. P. (1976): "On the Goldstein-Levitin-Polyak Gradient Projection Method" , IEEE Trans. on Automatic Control, V. AC-21, pp. 174 - 183.

Broyden, C. G. (1967): "Quasi-Newton Methods and their Application to Function Minimisation" , Math. Comp., V. 21, pp. 368 - 381.

Broyden, C. G. (1969): "A New Method for Solving Nonlinear Simultaneous Equations" , Comp. J., V. 12, pp. 95 - 100.

Broyden, C. G. (1970): "The Convergence of a Class of Double-Rank Minimization Algorithms 2 . The New Algorithm" , J. Inst. Maths. Applics., V. 6, pp. 222 - 231.

Broyden, C. G., Dennis, J. E. and Moré, J. J. (1973): "On the Local and Superlinear Convergence of Quasi-Newton Methods" , J. Inst. Maths. Applics., V. 12, pp. 223 - 245.

Caratheodory, C. (1967): "Calculus of Variations and Partial Differential Equations of First Order" , Holden-Day, San Francisco.

Davidon, W. C. (1959): "Variable Metric Method for Minimization" , Argonne Nat. Lab., Report ANL-5990 rev.

Demyanov, V. F. and Rubinov, A. M. (1967): "The Minimization of a Smooth Convex Functional on a Convex Set" , SIAM J. Control, V. 5, pp. 280 - 294.

Dennis, J. E. (1972): "On Some Methods Based on Broyden's Secant Approximation to the Hessian" , in: Numerical Methods for Nonlinear Optimisation, F. A. Lootsma (ed.), Academic Press, New York.

Dennis, J. E. and Moré, J. J. (1974): "A Characterisation of Superlinear Convergence and its Application to Quasi-Newton Methods" , Maths. Comp., V. 28, pp. 549 - 560.

Dennis, J. E. and Moré, J. J. (1977): "Quasi-Newton Methods, Motivation and Theory" , SIAM Review, V. 19, pp. 46 - 89.

Dixon, L. C. W. (1972): "Quasi-Newton Algorithms Generate Identical Points" , Math. Prog., V. 2, pp. 383 - 387.

Fletcher, R. (1970): "A New Approach to Variable Metric Algorithms" , Comp. J., V. 13, pp. 317 - 322.

Fletcher, R. (1972): "An Algorithm for Solving Linearly Constrained Optimization Problems" , Math. Prog., V. 2, pp. 133 - 165.

Fletcher, R. and Powell, M. J. D. (1963): "A Rapidly Convergent Descent Method for Minimization" , Comp. J., V. 6, pp. 163 - 168.

Gill, P. E. and Murray, W. (1972): "Two Methods for the Solution of Linearly Constrained and Unconstrained Optimization Problems" , Nat. Phys. Lab. Rept., NAC 25.

Gill, P. E. and Murray, W. (1973): "Quasi-Newton Methods for Linearly Constrained Optimization" , Nat. Phys. Lab. Rept., NAC 32.

Gill, P. E. and Murray, W. (1974,a): "Newton-Type Methods for Linearly Constrained Optimization" , in: Numerical Methods for Constrained Optimization, P. E. Gill and W. Murray (eds.), Academic Press, London.

Gill, P. E. and Murray, W. (1974,b): "Quasi-Newton Methods for Linearly Constrained Optimization" , in: Numerical Methods for Linearly Constrained Optimization, P. E. Gill and W. Murray (eds.), Academic Press, London.

Gill, P. E. and Murray, W. (1978): "Numerically Stable Methods for Quadratic Programming" , Math. Prog., V. 14, pp. 349 - 372.

Goldfarb, D. (1969): "Extension of Davidon's Variable Metric Algorithms to Maximization under Linear Inequality and Equality Constraints" , SIAM J. Appl. Math., V. 17, pp. 739 - 764.

Goldfarb, D. (1970): "A Family of Variable Metric Methods Derived by Variational Means" , Maths. Comp., V. 24, pp. 23 - 26.

Goldstein, A. A. (1964): "Convex Programming in Hilbert Space", Bull. AMS, V. 70, pp. 709 - 710.

Han, S.-P. (1976): "Superlinearly Convergent Variable Metric Algorithms for General Nonlinear Programming Problems", Math. Prog., V. 11, pp. 263 - 282.

Han, S.-P. (1977): "A Globally Convergent Method for Nonlinear Programming", JOTA, V. 22, pp. 297 - 309.

Householder, A. S. (1964): The Theory of Matrices in Numerical Analysis, Blaisdell, New York.

Huang, H. Y. (1970): "A Unified Approach to Quadratically Convergent Algorithms for Function Minimization", JOTA, V. 5, pp. 405 - 423.

Levitin, E. S. and Polyak, B. T. (1966): "Constrained Minimization Methods", USSR Comp. Math. and Math. Phys., V. 6, pp. 1 - 50.

Luenberger, D. G. (1969): Optimization by Vector Space Methods, J. Wiley, New York.

McCormick, G. P. and Tapia, R. A. (1972): "The Gradient Projection Method Under Mild Differentiability Conditions", SIAM J. Control, V. 10, pp. 93 - 98.

Moré, J. J. and Trangenstein, J. A. (1976): "On the Global Convergence of Broyden's Method", Math. of Comp., V. 30, pp. 523 - 540.

Murtagh, B. A. and Sargent, R. W. H. (1969): "A Constrained Minimization Method with Quadratic Convergence", in: Optimization, R. Fletcher (ed.), Academic Press, London and New York.

Ortega, J. M. and Rheinboldt, W. C. (1970): Iterative Solution of Nonlinear Equations in Several Variables, Academic Press, London and New York.

Palomares, U. M. G. and Mangasarian, O. L. (1974): "Superlinearly Convergent Quasi-Newton Algorithms for Nonlinearly Constrained Optimization Problems", Comp. Sciences Tech. Rep. No. 195, University of Wisconsin, Madison.

Polak, E. (1971): Computational Methods in Optimization, Academic Press, London and New York.

Powell, M. J. D. (1966): "Minimization of Functions of Several Variables" , in: Numerical Analysis: an Introduction, J. Walsh(ed.), Academic Press, London.

Powell, M. J. D. (1970.a): "A New Algorithm for Unconstrained Optimization", in: Nonlinear Programming, J. B. Rosen, O. L. Mangasarian, K. Ritter (eds.), Academic Press, New York.

Powell, M. J. D. (1970,b): "A Fortran Subroutine for Unconstrained Minimization Requiring First Derivatives of the Objective Functions" , A.E.R.E. Harwell Rept., R. 64 - 69.

Powell, M. J. D. (1974): "Unconstrained Minimization and Extensions for Constraints" , in: Mathematical Programming in Theory and Practice, P. L. Hammer and G. Zontendijk (eds.), North-Holland, Amsterdam.

Powell, M. J. D. (1975): "A View of Unconstrained Optimization" , A.E.R.E. Harwell Rept., C.S.S. 14.

Powell, M. J. D. (1977,a): "A Fast Algorithm for Nonlinearly Constrained Optimisation Calculations" , presented at the 1977 Dundee Conference on Numerical Analysis.

Powell, M. J. D. (1977,b): "The Convergence of Variable Metric Methods for Nonlinearly Constrained Optimization Calculations" , Argonne Nat. Lab., Report TM-315.

Psenichny, B. N. and Danilin, Y. M. (1978): Numerical Methods in Extremal Problems, Mir Publishers, Moscow.

Robinson, S. M. (1974): "Perturbed Kuhn-Tucker Points and Rates of Convergence for a Class od Nonlinear Programming Algorithms" , Math. Prog., V. 7, pp. 1 - 16.

Shanno, D. F. (1970): "Conditioning of Quasi-Newton Methods for Function Minimization" , Math. Comp., V. 24, pp. 647 - 654.

Rosen, J. B. (1960): "The Gradient Projection Method for Nonlinear Programming. Part I: Linear Constraints" , SIAM J. Appl. Math., V. 8, pp. 181 - 217.

Wilkinson, J. (1965): The Algebraic Eigenvalue Problem, Oxford University Press, Oxford.

Wilson, R. B. (1963): A Simplicial Algorithm for Concave Programming, Ph. D. Diss.,Grad. School of Business Administration, Harvard University.

Wolfe, P. (1969): "Convergence Conditions for Ascent Methods", SIAM Review, V. 11, pp. 226 - 235.

REFERENCES FOR CHAPTER 5

Bray, J. (1972): "Some Considerations in the Practice of Economic Management", PREM Discussion Paper No. 2.

Bray, J. (1975): "Optimal Control of a Noisy Economy with the U.K. as an Example", J. R. Statist. Soc., Series A, V. 138, Part 3.

Broyden, C. (1967): "Quasi-Newton Methods and their Application to Function Minimisation", Math. of Comp., V. 21, pp. 368-381.

Canon, M. D., Cullum, C. D. and Polak, E. (1970): Theory of Optimal Control and Mathematical Programming, McGray-Hill, New York.

Davidon, W. C. (1968): "Variance Algorithm for Minimization", Comp. J., V. 10, pp. 406-410.

Dennis, J. E. and More, J. J. (1977): "Quasi-Newton Methods, Motivation and Theory", SIAM Review, V. 19, pp. 47-89.

Fiacco, A. V. and McCormick, G. P. (1968): Nonlinear Programming: Sequential Unconstrained Minimization Techniques, John Wiley, New York.

Friedman, B. (1972): "Optimal Economic Stabilization Policy: An Extended Framework", J. Polit. Econ., V. LXXX, pp. 1002-1022.

Frisch, R. (1972): "Cooperation between Politicians and Econometricians on the Formalization of Political Preferences", University of Oslo, Institute of Economics Reprint Series No. 90.

Holly, S., Rustem, B., Zarrop, M., (1979): Optimal Control for Econometric Models, Macmillan, London.

Holly, S., Rustem, B., Westcott, J. H., Zarrop, M. B. and Becker, R. (1979): "Control Exercises with a Small Dynamic Linear Model of the U.K. Economy", in: Holly, Rustem, Zarrop (1979) above.

Kornai, J. (1975): Mathematical Planning of Structural Decisions, North Holland, Amsterdam.

Livesey, D. (1973): "Can Macro-Economic Planning Problems Ever be Treated as a Quadratic Regulator Problem", IEE Conference Publication, V. 101, pp. 1-14.

Livesey, D. (1976): "Feasible Directions in Economic Policy", presented at the European Meeting of the Econometric Society, Helsinki.

Luenberger, D. (1969): Optimization by Vector Space Methods, John Wiley, New York.

Murtagh, B. A. and Sargent, R. W. H. (1969): "A Constrained Minimization Method with Quadratic Convergence", in: Optimization, R. Fletcher (ed.), Academic Press, New York and London.

Pindyck, R. S. (1973): Optimal Planning for Economic Stabilization, North Holland, Amsterdam.

Polak, E. (1971): Computational Methods in Optimization, Academic Press, New York and London.

Powell, M. J. D. (1974): "Unconstrained Minimization and Extensions for Constraints", in: Mathematical Programming in Theory and Practice, P. L. Hammer and G. Zontendijk (eds.), North Holland, Amsterdam.

Sen, A. K. (1970): Collective Choice and Social Welfare, Holden-Day, San Francisco.

Westcott, J. H., Holly, S., Rustem, B. and Zarrop, M. (1976): "Control Theory in the Formulation of Economic Policy", Submission to the Committee of Enquiry on Policy Optimisation, PREM Discussion Paper No. 17, Department of Computing and Control, Imperial College.

REFERENCES FOR CHAPTER 6

Athans, M., Kuh, E., Ozkan, T., Papademos, L., Pindyck, R., Wall, K. (1976): "Sequential Open-loop Optimal Control of a Nonlinear Macroeconomic Model", in: Frontiers of Quantitative Economics, (ed.)M. D. Intriligator, North-Holland, Amsterdam.

Bray, J. (1975): "Optimal Control of a Noisy Economy with the U.K. as an Example", Journal of the Royal Statistical Society, Series A, V. 138, part 3.

Broyden, C. G. (1965): "A Class of Methods for Solving Nonlinear Simultaneous Equations", Mathematics of Computation, V . 25, pp. 223-245.

Cannon, M. D., Cullum, C. D., Polak, E. (1970): Theory of Optimal Control and Mathematical Programming, McGraw-Hill, New York.

Chow, G. C. (1975): Analysis and Control of Dynamic Economic Systems, John Wiley, New York.

Chow, G. C. (1979): "Effective Use of Econometric Models in Macroeconomic Policy Formulation", in: Holly, Rustem, Zarrop (1979) below.

Committee on Policy Optimisation (1978): Report, HMSO Cmnd. 7148.

Davis, M. H. A. (1977): Linear Estimation and Stochastic Control, Chapman and Hall, London.

Dennis, J.E. (1977): "Nonlinear Least Squares and Equations",in:The State of the Art in Numerical Analysis, (ed.)D.A.H. Jacobs, Academic Press, New York.

Gill, P.E., Golub, G.,Murray,W. and Saunders,M.A.(1974):"Methods for Modifying Matrix Factorizations", Math. Comp.,V.29,pp.1051-1077.

Gill, P.E. and Murray, W. (1972):"Quasi-Newton Methods for Unconstrained Optimization", J.Inst. Math. and Applics.,V.9, pp. 91-108.

Gill, P. E. and Murray, W. (1974,a): "Safeguarded Step-Length Algorithms for Optimization Using Descent Methods", National Physical Laboratory, Report NAC 37, Teddington, England.

Gill, P. E. and Murray, W. (1974,b): "Newton Type Methods for Linearly Constrained Optimization, in: Numerical Methods for Constrained Optimization, (eds.) P. E. Gill and Murray, Academic Press, London.

Gill, P. E. and Murray, W. (1976): "Algorithms for the Solution of the Nonlinear Least Squares Problem", National Physical Laboratory, Report NAC 71, Teddington, England.

Goldstein, A. (1967): Constructive Real Analysis, Harper and Row, London.

Goldfarb, D. (1976): "Factorized Variable Metric Methods for Unconstrained Optimization", Math. of Comp., V.30, pp.796-811.

Holbrook, R.S. (1974): "A Practical Method for Controlling a Large Nonlinear Stochastic System", A.E.S.M., V. 3, pp. 155-176.

Holly, S., Rustem, B., Zarrop, M. B. (eds.) (1979): Optimal Control for Econometric Models, Macmillan, London.

Holly, S., Rustem, B., Westcott, J. H., Zarrop, M. B. and Becker, R. (1979): "Control Exercises with a Small Linear Model of the U.K. Economy", in: Holly, Rustem, Zarrop (1979) above.

Karakitos, E., Rustem, B. and Zarrop, M. B. (1979): "Optimal Control and the Monetarist Controversy", PROPE Discussion Paper No.29, Department of Computing and Control, Imperial College, London.

Kendrick, D. A. and Majors, J. (1974): "Stochastic Control with Uncertain Macroeconomic Parameters", Automatica, V. 10, pp.587-594.

Klein, L. R. (1979): "Managing the Modern Economy: Econometric Specification", in: Holly, Rustem, Zarrop (1979) above.

Kowalik, J. and Osborne, M. R. (1968): Metnods for Unconstrained Optimization Problems, American Elsevier, New York.

London Business School Quarterly Econometric Nodel of the U.K. Economy, London Business School, (August 1975/January 1979).

Mantell, J. B. and Lasdon, L. S. (1978): A GRG Algorithm for Econometric Control Problems, Annals of Economic and Social Measurement, V. 1.6, pp.581-598.

Marvill, E. S. (1978): Exploiting Sparsity in Newton-Type Methods, Ph.D. Thesis, Cornell University.

Moré, J. J. and Trangenstein, J. A. (1976): "On the Global Convergence of Broyden's Method", Mathematics of Computation, V. 30, pp.525-540.

Ortega, J. M. and Rheinboldt, W. C. (1970): Iterative Solution of Nonlinear Equations in Several Variables, Academic Press, New York.

Polak, E. (1971): Computational Methods in Optimization, Academic Press, New York.

Preston, R. S., Klein, L. R. O'Brien, Y. C., Brown, B. W. (1976): "Control Theory Simulations Using the Wharton Long Term and Industry Forecasting Model", Wharton E.F.A.

Powell, M. J. D. (1970,a): "A Hybrid for Nonlinear Equations", in: Numerical Methods for Non-Linear Algebraic Equations, (ed.) P. Rabinowitz, Gordon and Breach, London.

Powell, M. J. D. (1970,b): "A Fortran Subroutine for Solving Systems of Nonlinear Algebraic Equations", in: Numerical Methods for Nonlinear Algebraic Equations, (ed.) P. Rabinowitz, Gordon and Breach, London.

Rustem, B., Velupillai, K. and Westcott, J. H. (1978): "Respecifying the Weighting Matrix of a Quadratic Objective Function", Automatica, V. 14, pp.567-582.

Rustem, B. (1980): "Policy Optimisation Algorithms for Nonlinear Econometric Models", in: Analysis and Optimisation of Systems, A. Bensoussan and J.L. Lions (eds.), Springer-Verlag, Berlin.

Rustem, B. and Zarrop, M. B. (1979,a): "A Newton Method for the Optimization and Control of Nonlinear Econometric Models", Journal of Economic Dynamics and Control, V. 1, pp.238-300.

Rustem, B. and Zarrop, M. B. (1979,b): "A Quasi-Newton Algorithm for the Control of Nonlinear Econometric Models", presented at the Conference on Economics and Control, Cambridge, U.K., June 18-21, forthcoming in Large Scale Systems.

Schubert, L. K. (1970): "Modification of a Quasi-Newton Method for Nonlinear Equations with a Sparse Jacobian", Mathematics of Computation, V. 24, pp. 27-30.

Vallet, E. A. (1977): Optimization of a Linear Economic Model , M.Sc. Thesis, Imperial College, London.

Wall, K. D. and Westcott, J. H. (1975): "Policy Optimization Studies with a Simple Control Model of the U.K. Economy", PREM Discussion Paper No.9, Imperial College, London.

REFERENCES FOR CHAPTER 7

Farah, J. L. (1977): Optimization of Stochastic Systems, Ph.D. Thesis, Imperial College, London.

Kornai, J. (1975): Mathematical Planning of Structural Decisions, North-Holland, Amsterdam.

Kushner, H. J. and Clark, D. S. (1978): Stochastic Approximation Methods for Constrained and Unconstrained Systems, Springer Verlag, New-York.

Parkinson, D. (1978): "High Speed Computing" , Phys. Bull., V. 29, pp. 464 - 465.

Polyak, B. T. (1978): "Nonlinear Programming in the Presence of Noise", Math.Prog., V. 14, pp. 87 - 97.

Rustem, B. and Velupillai, K. (1979): "A New Approach to the Bargaining Problem" , in: New Trends in Dynamic System Theory and Economics, M. Aoki and A. Marzollo (eds.), Academic Press, New-York and London.

Rustem, B. and Zarrop, M. B. (1980): "A Newton-Type Algorithm for a Class of N-Player Dynamic Games Using Nonlinear Econometric Models", Proceedings of the IFAC/IFORS Conference on Dynamic Modelling and Control of National Economics, Warsaw, to be published by Pergamon Press.

Theil, H. (1964): Optimal Decision Rules for Government and Industry, North-Holland, Amsterdam.

Lecture Notes in Control and Information Sciences

Edited by A. V. Balakrishnan and M. Thoma

Vol. 22: Optimization Techniques
Proceedings of the 9th IFIP Conference on
Optimization Techniques,
Warsaw, September 4–8, 1979
Part 1
Edited by K. Iracki, K. Malanowski, S. Walukiewicz
XVI, 569 pages. 1980

Vol. 23: Optimization Techniques
Proceedings of the 9th IFIP Conference on
Optimization Techniques,
Warsaw, September 4-8, 1979
Part 2
Edited by K. Iracki, K. Malanowski, S. Walukiewicz
XV, 621 pages. 1980

Vol. 24: Methods and Applications
in Adaptive Control
Proceedings of an International Symposium
Bochum, 1980
Edited by H. Unbehauen
VI, 309 pages. 1980

Vol. 25: Stochastic Differential Systems –
Filtering and Control
Proceedings of the IFIP-WG7/1 Working Conference
Vilnius, Lithuania, USSR, Aug. 28 – Sept. 2, 1978
Edited by B. Grigelionis
X, 362 pages. 1980

Vol. 26: D. L. Iglehart, G. S. Shedler
Regenerative Simulation of Response
Times in Networks of Queues
XII, 204 pages. 1980

Vol. 27: D. H. Jacobson, D. H. Martin, M. Pachter, T. Geveci
Extensions of Linear-Quadratic Control Theory
XI, 288 pages. 1980

Vol. 28: Analysis and Optimization of Systems
Proceedings of the Fourth International
Conference on Analysis and Optimization of Systems
Versailles, December 16–19, 1980
Edited by A. Bensoussan and J. L. Lions
XIV, 999 pages. 1980

Vol. 29: M. Vidyasagar,
Input-Output Analysis of Large-Scale
Interconnected Systems –
Decomposition, Well-Posedness and Stability
VI, 221 pages. 1981

Vol. 30: Optimization and Optimal Control
Proceedings of a Conference Held at
Oberwolfach, March 16–22, 1980
Edited by A. Auslender, W. Oettli, and J. Stoer
VIII, 254 pages. 1981

Vol. 31: Berc Rustem
Projection Methods in Constrained
Optimisation and Applications
to Optimal Policy Decisions
XV, 315 pages. 1981

